SpringerBriefs in Education

More information about this series at http://www.springer.com/series/8914

Alan Bain · Lucia Zundans-Fraser

Rising to the Challenge of Transforming Higher Education

Designing Universities for Learning and Teaching

 Springer

Alan Bain
Faculty of Education
Charles Sturt University
Bathurst, NSW
Australia

Lucia Zundans-Fraser
Faculty of Education
Charles Sturt University
Bathurst, NSW
Australia

ISSN 2211-1921   ISSN 2211-193X   (electronic)
SpringerBriefs in Education
ISBN 978-981-10-0259-5   ISBN 978-981-10-0261-8   (eBook)
DOI 10.1007/978-981-10-0261-8

Library of Congress Control Number: 2015957236

Printed on acid-free paper

This Springer imprint is published by SpringerNature
The registered company is Springer Science+Business Media Singapore Pte Ltd.

Preface

Serious and fundamental questions are being posed about the quality of learning and teaching in universities. These questions relate to rapid change in the way learning and teaching is delivered, the cost and quality of courseware, responsiveness to learner heterogeneity, and faculty professionalization (Bokor 2012; Bradley et al. 2008; Norton et al. 2013). The direction universities take in answering these questions has consequences for their design, their staffing, the way they create and deliver their learning and teaching programs, and the way they compete with each other. According to the proponents of transformational change in higher education, the future services offered by universities will exist in a national and global marketplace instead of a state and regional one, and more often in a virtual as opposed to bricks and mortar delivery model (Harden 2012). While the drivers for change in the university sector and associated debates have existed for some time, there is an undoubted increasing cadence in the discussion and a demonstrable sense of fear driving a call for action within and among universities. Increasingly, cost, competitiveness, effectiveness, and viability are linked to responsiveness to student needs, the quality of the curriculum, the learning experience, and student outcomes (Kuh et al. 2015). The result is a heightened focus on how universities are organized to deliver those outcomes.

One of the ironies of change initiatives in universities is they frequently lack a coherent theory and research to practice framework for such important work, which is ironical, given that universities are the places where theories are created and applied in so many domains. There is also a concern that the scale of change is underwhelming, often limited to projects focused directly on improving aspects of learning and teaching without undertaking the institution-wide structural reform (i.e., an integrated approach to the technology, the governance, and personnel reforms) necessary to support those initiatives and increase the likelihood that they are both sustainable and scalable. Such projects are frequently positioned as transformational although their scope and depth rarely line up with the requirements of transformational change.

The purpose of this brief is to share an approach to change that extends beyond traditional strategic planning or scope–limited, project-based approaches and includes examples of the ways in which theory and research can guide practice in transformational organizational change in higher education.

References

Bokor, J. (2012). University of the future: A thousand year old industry on the cusp of profound change. Australia: Ernst & Young.

Bradley, D., Noonan, P., Nugent, H., & Scales, B. (2008). Review of Australian higher education. Retrieved from http://www.innovation.gov.au/HigherEducation/Documents/Review/PDF/Higher%20Education%20Review_one%20document_02.pdf

Harden, N. (2012). The end of the university as we know it. Retrieved from http://www.the-american-interest.com/articles/2012/12/11/the-end-of-the-university-as-we-know-it/

Kuh, G., Ikenberry, S., Jankowski, N., Cain, T., Ewell, P., Hutchings, P., & Kinzie, J. (2015). Using evidence of student learning to improve higher education. San Francisco: Jossey-Bass.

Norton, A., Sonnemann, J., & Cherastidtham, I. (2013). Taking university teaching seriously. Grattan Institute.

Contents

Chapter 1
The Self Organizing University (SOU)

Abstract This chapter introduces the self-organizing university model including a description of its elements, rationale, theoretical base, underpinning learning and teaching research, and the role of technology. The chapter also describes the outcomes that can be expected from using the model as a change approach in higher education.

Keywords Self-organization · Learning analytics · Theory · Research · Learning and teaching model · Scale · Transformational change · Technology system

The brief describes the theory and research literature that underpin the self-organizing university (SOU), a transformational change model for learning and teaching in higher education. It includes a description of the curricular context and need for the approach and the way multiple bodies of literature intersect in the theory, design and practice that drive the SOU approach in five areas. They are: context, theoretical framework, learning and teaching research, efficacy research and technology. Examples of the way each of the elements contribute to the model are included throughout. The context for SOU examines the following:

1. The existing state of program design, development and reform in higher education.
2. The theoretical basis for SOU and the literature associated with that theory.
3. The learning and teaching research that provides content for the SOU.
4. The efficacy research applying the theory in K-12 and higher education.
5. The approach taken to the design, including technologies for transformational change.

1.1 What Is the SOU?

The SOU is a multi-year transformational organizational change model for transforming learning and teaching in higher education contexts. This includes:

© The Author(s) 2016   1
A. Bain and L. Zundans-Fraser, *Rising to the Challenge of Transforming
Higher Education*, SpringerBriefs in Education,
DOI 10.1007/978-981-10-0261-8_1

- Building a new way of designing, developing, implementing and evaluating programs.
- Creating a software system for learning design, development, implementation and evaluation.
- Developing a new regulatory and approval process for securing and maintaining programs of the highest quality.
- Developing a new organizational design to support the learning and teaching model including:

  – Employing collaborative process to embed the approach in the normal work of the university.
  – Embedding the model in the role descriptions of key learning and teaching staff.
  – Embedding the model in the promotion process.
  – Embedding the model in the performance management process.

- The development of a new teaching and learning analytics system that can both drive and integrate with the broader university strategy.

The SOU approach is a model for universities committed to a transformational level of organizational change commensurate with the learning and teaching challenges they face. It includes developing the technologies, systems, organizational design and implementation methodology required to transform their learning and teaching capability at scale.

1.2   What Are the Expected Outcomes of the SOU?

- A new model and delivery approach for online and blended learning.
- Research-based program design and implementation at scale based upon the implementation of the learning and teaching model.
- Improved workforce capacity in learning and teaching.
- An organizational structure and regulatory process for program design, implementation and evaluation focused on evidence of learning and teaching quality.
- A high-value system of feedback and analytics that provides a university with the capacity for continuous change and improvement necessary to sustain leadership in learning and teaching.
- A self-organizing technology system and analytics model.

These expected outcomes constitute the learning and teaching deliverables required to meet and sustain a genuinely distinctive approach to learning and teaching at the scale of the whole university.

Chapter 2
The State of Program Design and Development

Abstract Why take on a whole-of-university approach to the improvement of learning and teaching? This chapter answers this question by describing the state of current practice followed by a rationale for the self-organizing university model including the need for theory, collaboration, and new institutional practice for governing the learning and teaching process.

Keywords Collaboration · Theory · Research to practice gap · Curriculum change · Climate · Institutional practice · Autonomy · Learning and teaching

The climate of change and uncertainty in universities has heightened the focus on learning and teaching. This includes concern about entry and exit standards, participation (Bradley et al. 2008), issues of equity and inclusion (Devlin et al. 2012), and the impact of accountability (Kimber and Ehrich 2015). The need to address these concerns is reflected in a range of initiatives that include projects to track and predict student performance and success, the innovative use of technology, and external benchmarking (Kemp and Norton 2014), the use of outcomes-based education practices (Barman et al. 2014; Lam and Tsui 2014), program evaluations (Ronsholdt and Brohus 2014), enhanced student perceptions of learning experiences (Kuhn and Rundle-Thiele 2009) and the measurement of student satisfaction and performance.

According to Norton et al. (2013), the need for good teaching in universities has never been greater. Good teaching begins with a strong responsive curriculum. Bradley et al. (2008) challenged universities to view increased participation in higher education as a driver for curriculum reform. With current curriculum reform promoting the importance of catering for the needs of individual students across all disciplines, changes in the design and delivery of programs are also necessary.

What is not evident internationally at this time are coherent and integrated approaches to the broad-based fundamental transformation of universities to address the tensions and challenges associated with becoming better learning organizations. This includes the simultaneous transformation of policy, work practice, organizational design, technology, and support around learning and

© The Author(s) 2016   3
A. Bain and L. Zundans-Fraser, *Rising to the Challenge of Transforming Higher Education*, SpringerBriefs in Education,
DOI 10.1007/978-981-10-0261-8_2

teaching that are commensurate with the order of magnitude of the challenges facing universities. How university programs are designed has not been addressed through reform, although traditional design methods have been widely criticised (Florian et al. 2010; Forlin 2010; Song-Turner and Willis 2011).

Four areas of need have been consistently identified in the higher education literature related to program design and development. These are: the need for a theoretical basis for program reform and design (Goldspink 2007; Levin 2010); the need to close the theory-to-practice gap between (a) research in higher education, and (b) what is done in practice (Hora 2012; Norton et al. 2013; Song-Turner and Willis 2011); the need to utilise collaborative practice, including the way groups of people work together to achieve common goals (Burgess 2004; Furlonger et al. 2010; Oliver and Hyun 2011; Winn and Zundans 2004); and the impact of prevailing institutional practice on program reform, design and innovation (Bradley et al. 2008; Levin 2010; Oliver 2013; Sabri 2010).

2.1 Need for Theory

A number of researchers have identified the need for curriculum innovation to be theorised beyond actual curriculum content (Cochran-Smith and Zeichner 2005; Hoban 2004; Kezar and Lester 2009). This refers to the kind of theory that links design process to the full scope of the change required by organizations to engage in substantive and sustainable curriculum reform. Such theory addresses the way the organization is structured for learning and teaching, as well as the content and process specifically related to the design of curricula. It is clear that unifying theory is necessary to bring these interdependent aspects together.

2.2 Research to Practice Gap

There is also a gap between research and practice with the most prevalent explanation suggesting that researchers and practitioners operate within quite different cultures (Kezar 2000; Schalock et al. 2006). Kezar (2000) suggests that the dichotomy between theory and practice is a social construction. Within the institutional context, theory is broadly constructed to mean work that is done in research, principally in university settings, that in turn is expected to impact upon the field. However, this fundamental operating assumption of universities has not been applied to the field of curriculum learning and teaching in higher education where innovations are rarely theorised and only loosely connected to related research.

Multiple factors come into play at the higher education level that continue to separate theory from practice—the culture of the institution, socialisation of the faculty, the discipline foci of the institution, and the continued tension within higher education itself as a professional identity (Kezar and Lester 2009). This suggests that addressing the theory to practice gap requires high-level and broad-based change capable of engendering new cultures of practice. Those new cultures need to address the separation from practice (Grima-Farrell 2012) and reconcile the acquisition of practical knowledge with traditional theoretical knowledge (Berry 2007). From a SOU perspective, the magnitude of the gap in all sectors of the education field is indicative of a field that is pre-paradigmatic in its development and evolution (Kuhn 1996). The reconciliation of theory and practice is necessary for any theory to be complete and for practice to be informed.

2.3 Collaboration

Successful and long-term change requires collaboration, yet higher education is often a solitary and isolating profession, with university structures separating disciplines and individuals within disciplines. Kezar and Lester (2009) described the conditions required for a higher education institution to support collaboration. They found that teaching, research, governance and management all needed to be altered to support the collaborative process. Collaboration, and the process required for effective collaboration, can be viewed at several levels: collaboration at the institutional level where academics work together; academics in a specialist field working together to create and develop new program content; collaboration in the field; and collaboration between educators in different contexts, such as between academics and teachers (Zundans-Fraser 2014). Researchers indicate that despite recognising the benefits of collaboration, teachers in universities continue to work largely in isolation (Norton et al. 2013; Zundans-Fraser and Bain 2015a).

2.4 Institutional Practice

Knapp and Brandon (1998), suggest that the institutional practice of universities also impedes curricular reform. The authors identify four prominent features of university structure and institutional practice that impede innovation. They are: The knowledge-centric nature of the organization, position hierarchy, promotion systems and departmental autonomy. These factors interact in ways that perpetuate the status quo by making collaboration and organizational coherence difficult. They create insurmountable challenges for academics attempting anything a little different and innovative while perpetuating "private practice" work environments that inhibit collaboration. In their study, Zundans-Fraser and Bain (2015a) found that academics believed working in a collaborative manner led to an open and transparent design process, the valuing of different opinions and expertise, shared knowledge and responsibility, and ultimately the design of a quality program. Most of the challenges identified by the academics had to do with institutional practices where the investment of time and recognition of the work done were lacking. This meant that although the academics had shared ownership of the program and valued collaborative practice, working in such a manner with existing processes was challenging. An institutional process that embeds collaboration within organizational structures as a standard expectation was suggested as a way to move institutional program design and review processes from a data entry exercise to one focused on quality design.

When a level of flexibility is sought against centralised control within this complex environment (Knight 2001; Macdonald 2003), the enthusiasm for self-organization demonstrated by educators is dampened at administrative levels due to concern about potential risks that may never eventuate (Goldspink 2007); and when the greatest focus is on rationalisation and efficiency (Coate and Tooher 2010). These concerns reflect the tension that exists between the organizational structure and realities of higher education institutions and their daily operations. The everyday restrictions and constraints of higher education practice, such as tight deadlines and expansive documentation requirements, appear to work against curriculum coherence and depth. The institutional practice of many institutions focuses on documentation requirements that are not related to the quality of learning and teaching (Zundans-Fraser 2014) and where ideas of quality are shaped by the extant demands of the organization around the way curriculum is developed and documented (Borko et al. 2006; Tudor 2006). Curriculum reform and design that goes beyond the regular institutional cycles of student evaluation, staff reflection, and the selection and design of new program materials is particularly challenging, as it requires time and dedicated participation that some researchers claim are not a natural part of higher education culture (Burgess 2004; Oliver and Hyun 2011).

2.5 Summary

This overview of key program development issues provides an evidence base for the claims made in the introduction to this brief. Reforming curriculum design and delivery involves transformational change in the way universities are structured and organised. Existing processes are not only frequently incompatible with change and better practice, but can actively undermine such change through diffusion of purpose, poor alignment of goals and process, and contradictory structures for reward and recognition. Even though there has been a dramatic increase in educational reform efforts over the past 25 years, it is having insufficient impact to drive sustainable and scalable change (Hopkins and Levin 2000). Many academics instrumental in curriculum reform are open to innovations, change, and enhancing teaching and learning. However, reform often gets in the way by focusing on operational aspects of the system—governance, finance, workforce and accountability that do not support curricular innovation. A key contention of SOU is that curricular reform needs to drive structural improvement informed by complete theory to practice models of institutional change for better learning and teaching.

Fundamental change in the way university services are funded, designed and delivered, and the needs and characteristics of the students who pay for and receive the service are driving reform and change in universities that frequently lack the theoretical frameworks, institutional process, practice and collaborative cultures required to address the profound changes they face. An integrated program design approach, a more rigorous theoretical base for program reform and tight coherence between course work and practical work are required (Darling-Hammond 2006; Darling-Hammond et al. 2005). The type of reform that is advocated is in its infancy in the higher education sector, with little attention given to the evolution of program content or the way it can transform and renew an institution, although this situation is slowly changing (Oliver and Hyun 2011). Reform that moves beyond the regular institutional cycles of student evaluation, staff reflection, selection and design of new materials at the course level is particularly challenging as it requires time and dedicated participation that some researchers claim are not a natural part of higher education culture (Burgess 2004, Kezar and Lester 2009). An integrated program design approach, a more rigorous theoretical base for program reform and tight coherence between course-work and practice is required (Zundans-Fraser and Bain 2015b). Any change initiatives designed to address the key areas of need identified here need to be grounded in changes in the extant institutional requirements for program design.

References

Barman, L., Bolander-Laksov, K., & Silen, C. (2014). Policy enacted—Teachers' approaches to an outcome-based framework for course design. *Teaching in Higher Education, 19*(7), 735–746. doi:10.1080/13562517.2014.934346

Berry, A. (2007). *Tensions in teaching about teaching: Understanding practice as a teacher educator*. Dordrecht: Springer.

Borko, H., Liston, D., & Whitcomb, J. A. (2006). A conversation of many voices: Critiques and visions of teacher education. *Journal of Teacher Education, 57*(3), 199–204. doi:10.1177/0022487106287978

Bradley, D., Noonan, P., Nugent, H., & Scales, B. (2008). *Review of Australian higher education*. Retrieved from http://www.innovation.gov.au/HigherEducation/Documents/Review/PDF/Higher%20Education%20Review_one%20document_02.pdf

Burgess, H. (2004). Redesigning the curriculum for social work education: Complexity, conformity, chaos, creativity, collaboration? *Social Work Education, 23*(2), 163–183. doi:10.1080/0261547042000209189

Coate, K., & Tooher, M. (2010). The Galway symposium on design for learning: Curriculum and assessment in higher education. *Teaching in Higher Education, 15*(3), 347-354. doi:10.1080/13562511003740924

Cochran-Smith, M., & Zeichner, K. M. (2005). *Studying teacher education: The report of the AERA panel on research and teacher education*. Washington, DC: American Educational Research Association.

Darling-Hammond, L. (2006). Constructing 21st-century teacher education. *Journal of Teacher Education, 57*(3), 300–314. doi:10.1177/0022487105285962

Darling-Hammond, L., Hammerness, K., Grossman, P., Rust, F., & Shulman, L. (2005). The design of teacher education programs. In L. Darling-Hammond, J. Bransford & National

Academy of Education (Eds.), *Preparing teachers for a changing world: What teachers should learn and be able to do*. San Francisco, CA: Jossey-Bass.

Devlin, M., Kift, S., Nelson, K., & Smith, L. (2012). *Effective teaching and support of students from low socio-economic backgrounds: Resources for Australian Higher Education*. Retrieved from http://www.olt.gov.au/project-effective-teaching-and-support-students-low-socioeconomic-backgrounds-resources-australian-h

Florian, L., Young, K., & Rouse, M. (2010). Preparing teachers for inclusive and diverse educational environments: Studying curricular reform in an initial teacher education course. *International Journal of Inclusive Education 14*(7), 709–722. doi:10.1080/13603111003778536

Forlin, C. (2010). Teacher education reform for enhancing teachers' preparedness for inclusion. *International Journal of Inclusive Education, 14*(7), 649–653. doi:10.1080/13603111003778353

Furlonger, B. E., Sharma, U., Moore, D. W., & Smyth-King, B. S. (2010). A new approach to training teachers to meet the diverse learning needs of deaf and hard-of-hearing children within inclusive Australian schools. *International Journal of Inclusive Education, 14*(3), 289–308. doi: 10.1080/13603110802504549

Goldspink, C. (2007). Transforming education: Evidential support for a complex systems approach. *Emergence: Complexity & Organization, 9*(1/2), 77–92.

Grima-Farrell, C. (2012). *Identifying factors that bridge the research-to-practice gap in inclusive education: An analysis of six case studies*. (Unpublished doctoral dissertation). Charles Sturt University, Bathurst.

Hoban, G. F. (2004). Seeking quality in teacher education design: A four-dimensional approach. *Australian Journal of Education, 48*(2), 117–133. doi:10.1177/000494410404800203

Hopkins, D., & Levin, B. (2000). Educational reform and school improvement. *NIRA Review, 21–26*.

Hora, M. T. (2012). Organizational factors and instructional decision-making: A cognitive perspective. *The Review of Higher Education, 35*(2), 207–235. doi:10.1353/rhe.2012.0001

Kemp, D., & Norton, A. (2014). *Review of the demand driven funding system*. Retrieved from https://docs.education.gov.au/system/files/doc/other/review_of_the_demand_driven_funding_system_report_for_the_website.pdf

Kezar, A. (2000). Understanding the research-to-practice gap: A national study of researchers' and practitioners' perspectives. *New Directions for Higher Education, 110*, 9–19. doi:10.1002/he.11001

Kezar, A., & Lester, J. (2009). *Organizing higher education for collaboration: A guide for campus leaders*. San Francisco, CA: Wiley.

Kimber, M., & Ehrich, L. C. (2015). Are Australia's universities in deficit? A tale of generic managers, audit culture and casualisation. *Journal of Higher Education Policy and Management, 37*(1), 83–97. doi:10.1080/1360080X.2014.991535

Knapp, M. S., & Brandon, R. N. (1998). Building collaborative programs in universities. In M. S. Knapp (Ed.), *Paths to partnership: University and community as learners in inter professional education*. Oxford: Rowman & Littlefield Publishers.

Knight, P. T. (2001). Complexity and curriculum: A process approach to curriculum-making. *Teaching in Higher Education, 6*(3), 369–381. doi:10.1080/13562510120061223

Kuhn, T. (1996). *The structure of scientific revolutions* (3rd ed.). Chicago: University of Chicago.

Kuhn, K.-A. L., & Rundle-Thiele, S. R. (2009). Curriculum alignment: Exploring student perception of learning achievement measures. *International Journal of Teaching and Learning in Higher Education, 21*(3), 351–361.

Lam, B. H., & Tsui, K. T. (2014). Curriculum mapping as deliberation—Examining the alignment of subject learning outcomes and course curricula. *Studies in Higher Education*. doi:10.1080/03075079.2014.968539

Levin, B. (2010). Governments and education reform: Some lessons from the last 50 years. *Journal of Education Policy, 25*(6), 739–747. doi:10.1080/02680939.2010.523793

Macdonald, D. (2003). Rich task implementation: Modernism meets postmodernism. *Discourse: Studies in the Cultural Politics of Education, 24*(2), 247–262.

Norton, A., Sonnemann, J., & Cherastidtham, I. (2013). *Taking university teaching seriously.* Grattan Institute.

Oliver, B. (2013). Graduate attributes as a focus for institution-wide curriculum renewal: Innovations and challenges. *Higher Education Research & Development, 32*(3), 450–463.

Oliver, S. L., & Hyun, E. (2011). Comprehensive curriculum reform in higher education: Collaborative engagement of faculty and administrators. *Journal of Case Studies in Education, 2*, 1–20.

Ronsholdt, B., & Brohus, H. (2014). Towards more efficient student course evaluations for use at management level. *Tertiary Education and Management, 20*(1), 72–83. doi:10.1080/13583883.2014.881912

Sabri, D. (2010). Absence of the academic from higher education policy. *Journal of Education Policy, 25*(2), 191–205.

Schalock, H. D., Schalock, M. D., & Ayres, R. (2006). Scaling up research in teacher education: New demands on theory, measurement and design. *Journal of Teacher Education 57*(2), 102–119. doi:10.1177/0022487105285615

Song-Turner, H., & Willis, M. (2011). Re-engineering the course design and delivery of Australian tertiary education programmes: Perspectives from Chinese students. *Journal of Higher Education Policy and Management, 33*(5), 537–552. doi:10.1080/1360080X.2011.605228

Tudor, I. (2006). Teacher training and 'quality' in higher education language teaching: Strategies and options. *European Journal of Teacher Education, 29*(4), 519–532. doi:10.1080/02619760600944811

Winn, S., & Zundans, L. (2004). University and school connections: Enhancing literacy development of primary aged children. *Special Education Perspectives, 13*(1), 49–62.

Zundans-Fraser, L. (2014). *Self-organisation in course design: A collaborative, theory-based approach to course development in inclusive education.* (Unpublished doctoral dissertation). Charles Sturt University, Bathurst.

Zundans-Fraser, L., & Bain, A. (2015a). The role of collaboration in a comprehensive programme design process in inclusive education. *International Journal of Inclusive Education.* Advance online publication. doi:10.1080/13603116.2015.1075610

Zundans-Fraser, L., & Bain, A. (2015b). *How do institutional practices for program design and review address areas of need in higher education.* Manuscript submitted for publication.

Chapter 3
The SOU Model

Abstract In this chapter, the three assumptions underpinning the self-organizing university model are described along with a diagrammatic representation of the way the elements of the model interact to produce better learning and teaching outcomes.

Keywords Transformational change · Scale · Student experience · Self-organizing university · Curricular reform · Higher education

The SOU is based on three key assumptions that relate to the current drivers and context within higher education. They are:

1. Universities must make profound improvement in the universal design and delivery of all learning and teaching to respond to increasing student diversity.
2. The improvement must be demonstrable at scale. Viability will be linked to the quality of the individual student experience at scale.
3. Realising such improvement will require transformational change in the way universities are constituted for learning and teaching.

Transformational change is synonymous with addressing the needs identified in the literature related to curricular reform in higher education.

The SOU model was conceived in order to produce better programs and courses, greater student engagement, increased market share—especially in an online environment, market leading technology tools, and generate brand leverage through differentiation and leadership in the field. SOU assumes transforming learning and teaching occurs at a whole of university scale. To do so requires an interaction of theory, research, organizational design and process in a complete model. The model should create clear line of sight from theory to day-to-day practice.

Figure 3.1 is a high-level representation of the way theory, research design and implementation interact in the SOU. The diagram highlights an intentional,

© The Author(s) 2016
A. Bain and L. Zundans-Fraser, *Rising to the Challenge of Transforming Higher Education*, SpringerBriefs in Education,
DOI 10.1007/978-981-10-0261-8_3

Fig. 3.1 The SOU theory,
research, design and
implementation process

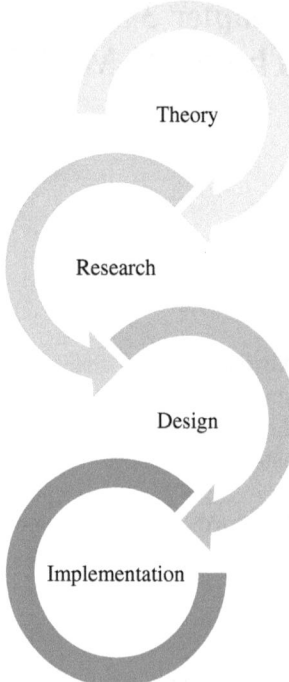

dynamic relationship between theory, research, design and implementation. The
subsequent sections of the brief describe the content in each area and the interaction
that defines the approach.

Chapter 4
Theory

Abstract Kurt Lewin said, "There is nothing so practical as a good theory" (Cherry in Selected quotations by Kurt Lewin, 2014). In this chapter a practical theory of the self-organizing university is introduced and described. The chapter shows how six theoretical principles can be translated into a practical model for change. The model empowers the agency that exists in the organization and provides specific guidance for transforming learning and teaching.

Keywords Self-organization · Simple rules · Embedded design · Self-similarity · Dispersed control · Emergent feedback · Schema · Emergence

Universities are often criticised for a failure to apply theory to the design and conduct of their own activities (Allen and Wright 2014; Coate and Tooher 2010; Fitzmaurice 2010). Kurt Lewin said, "There is nothing so practical as a good theory" (Cherry 2014). From a SOU perspective, theory is more than a set of ideas used to account for a phenomenon of interest. According to Kuhn (1996), the explanatory value of a theory stands at the confluence of its concepts, beliefs and the methods, systems and tools employed to articulate those concepts and beliefs in practice. In SOU, theory is a practical tool, used for model building and articulating the way things happen. In a SOU approach, theory should generate models of practice that influence the essential function of an organization or system not just the way it articulates intents or the way it is managed. Getting to scale with an innovation is also related to the role of theory. The efficacy of a theory can be seen as an expression of its capacity to generate a scalable solution to a problem or need in ways that resolve more issues than prevailing theories or approaches (Kuhn 1996).

The theory of self-organization and self-organizing systems (Juarrero and Rubino 2010; Kauffman 1995; Prigogine and Stengers 1984; Prokopenko 2013; Waldrop 1993) underpins the SOU approach. Self-organization is widely documented as an explanatory framework for systems in nature (Juarrero and Rubino 2010; Prigogine and Stengers 1984) and those involving human action and intervention including cities (Johnson 2001), economies (Krugman 1996), businesses (Pascale et al. 2000), schools (Bain 2007) and the design of technology systems

© The Author(s) 2016

A. Bain and L. Zundans-Fraser, *Rising to the Challenge of Transforming Higher Education*, SpringerBriefs in Education, DOI 10.1007/978-981-10-0261-8_4

(Bain and Weston 2012). Self-organization helps us to understand the non-linear and complex function of systems (Merry 1995). It indicates that systems are capable of powerful bottom-up change when those systems share a schema or framework for action (Gell-Mann 1994). With a shared schema, the system can disperse control to those agents. Empowered by their feedback and self-organizing activity, the system can succeed, grow, adapt and change without constant top-down intervention. Theories of self-organizing systems have generated great interest in many fields including education because they provide metaphors for change and adaptation without the constant top-down intervention associated with traditional management models (Scott 1991). They have a particular allure for education from two perspectives. First, they are frequently used to explain the complexity and often chaotic state of educational organizations (Morrison 2002) and second, they offer the possibility of better ways to share feedback, disperse control, distribute leadership, and innovate in those systems. In the SOU, the role and contribution of self-organization extends beyond the metaphorical, to the application of six design principles extracted from accounts and descriptions of self-organizing systems, to inform the design of educational organizations (Bain 2007). These six design principles are as follows.

4.1 Simple Rules or Commitments

In self-organizing systems remarkably complex behaviour can be stimulated by relatively simple rules (Bain 2007; Seel 2000; Sull and Eisenhardt 2012). In a higher education system, simple rules can be a commitment to a set of graduate attributes, to collaboration and shared practice, to authenticity in assessment, to evidence in pedagogy, and the role of feedback and transparency in program and course development processes. The development of these commitments is a dynamic, negotiated process. Organizations frequently identify commitments through strategic planning processes. What distinguishes a self-organizing system is the way those commitments are intentionally translated into the structure and normal work of the organization. The development of simple rules or commitments is a first step in the baseline and design phase of SOU, where the organization takes up what it believes and values by looking at its current needs and future aspirations in relation to those commitments. Simple rules provide a term of reference for successful collaboration and guidance about the constitution of groups within human systems. They enable agents within systems to work together successfully to build capacity with a common or shared understanding of their normal work and to enact a common schema. These simple rules are not mission statements; rather they are drivers for the form and function of a system. It is also important to note that in successful systems these rules are not extensive and cumbersome. They are easily understood by all agents within the system and become a powerful force for self-organisation.

By identifying and formulating simple rules within the planning process, all agents involved determine which concepts are relevant in an open and transparent manner. Agents can also emphasise particular areas of need or weakness as a focus. For example, the literature clearly demonstrates that collaboration is one such area (Lester and Kezar 2012; Levine and Marcus 2010; Marshall et al. 2011). Collaborative practice can be easily incorporated as a simple rule to ensure that all agents are mindful of that requirement throughout a course design process. Existing research (Chao et al. 2010; Perry 2014; Zundans-Fraser and Bain 2015a) shows that the use of a collaborative methodology increases the quality of a course in design and implementation. Specific teaching and learning competencies relevant to an individual program or course can also be incorporated and targeted through the course design process (e.g., criterion-based assessment or cooperative learning). The ultimate benefit is that, through the establishment of simple rules, the course or program design process is grounded in a common set of educational purposes, themes and assumptions (Lynch 2012; Tom 1997).

4.2 Embedded Design

This design principle involves thinking about and acting upon the ways simple rules can be enacted in the organization's design. This is all too often the missing piece in change initiatives. Frequently strategic planning identifies new ways of thinking and doing that are targeted for implementation in an existing organizational design that was created to do the very things that need to change and as a result perpetuate current practice. This was highlighted in the discussion of current issues in curriculum development. For example, most organizations would like to be more collaborative. However, their organizational structures reinforce private practice. It should come as no surprise that turning over a goal of increased collaboration to an organizational design that is non-collaborative is unlikely to produce successful operational outcomes. Without change to the form of the organization, a strategic objective no matter how well articulated will falter when the organization does not have the design required for effective implementation.

Embedded design involves developing the systems, tools and strategies that complete a theory based upon an organization's commitments (Kuhn 1996). In a higher education setting, this involves embedding collaborative process in the governance structure, creating tools that assist in sharing feedback, creating a policy about authentic assessment, or articulating processes for embedding graduate attributes in course and program design (Zundans-Fraser and Lancaster 2012). In all cases, embedded design means making operational in design those things that the organization has committed to.

Embedded design extends the traditional development trajectory for change in universities that frequently involve committee work, document and professional development in discrete areas to a comprehensive integrated design process. Documents are developed to reflect policies, and processes that in turn are embedded in role descriptions, technologies, feedback mechanisms and career paths. Commitments are embedded in all facets of the university and echo in each.

Importantly all of the features of the design can be seen in each element—in role descriptions, governance models, career trajectory, feedback mechanisms and the technologies used to engage with the work of the organization. The result is a design that possesses line of sight from its big ideas and commitments to day-to-day normal work. This draft design is tested in the pilot phase of the SOU process.

Embedded design creates self-repeating patterns (Waldrop 1993) by expressing simple rules in organizational design and by embedding these design features in all others. "It allows the beliefs and ideas of a theory to be connected to its essential systems and practices" (Bain 2007, p. 50). The principle of embedded design also creates a level of predictability among the various components within a system. In an educational context a commitment to collaboration can guide the way meetings happen across an organization and the roles of those who work in the organization (Zundans-Fraser and Bain 2015b). Embedded design ensures the system expresses its simple rules by embedding them at all levels of a group's activity (Zundans-Fraser and Lancaster 2012). When these simple rules are evident at all levels of the design and used in some way through day-to-day activities, capacity and familiarity are built that ensure self-reinforcement. An organization with an embedded design possesses a virtual form of agency because of the way it is constituted (Levy 2001). The practical expression of simple rules, the connections and self-repeating patterns empower those who engage with it. Working in the organization builds capacity with its intents and purposes because its systems, processes and tools express those intents and purposes and help to build capacity with them.

4.3 Similarity at Scale

Self-organizing systems tend to be flatter and less hierarchical than others, although they do possess levels and reflect differential sources of influence in hierarchical ways (Waldrop 1993). Similarity at scale is what happens when the key features of a system are embedded at all levels, making a system similar to itself (Gleick 2008; Merry 1995). This similarity is evident at different levels or scales regardless of the manner in which control is dispersed. In a university setting, this can mean that the way an executive team functions to approve and monitor the performance of the institutional program profile is focused on the same process and feedback as a faculty team or the team designing and implementing an individual program. This self-similarity is achieved by ensuring that the tools, systems and methods that operationalize the theory are similar at all levels of the organization. This principle is challenging for higher education organizations that tend to segment leadership and management and operational activity from strategic activity and by level in the organizational hierarchy. In a successful self-organizing system, the community "pays attention at scale" meaning that leaders are focused on the same things as those engaged in the day-to-day work in equally dynamic ways, just from a different elevation or perspective. Part of the SOU design and implementation process is to ensure that there is vertical and horizontal self-similarity in the process. When a program team meets to problem solve an issue with assessment or alignment they employ the same process and foci as a faculty or university level team looking at faculty or university performance in the same areas. The role of leadership and faculty is not segmented by strategic and operational responsibility. The organization maintains a united focus on learning and teaching at all levels.

4.4 Emergent Feedback

According to Pascale et al. (2000), feedback is the way a self-organizing system talks to itself. It is the way agency is amplified as the source of dynamic change and adaptation. Adaptive complex systems possess feedback mechanisms that enable all interactions to be fed back through iterative feedback loops (Andrews et al. 2012; Gleick 2008). A system that can feed back and forward as part of a network of constant exchange among individuals and groups can create emergent change in the organization. Feedback is termed emergent in SOU (Bain 2007; Bain and Swan 2011) because it emerges bottom-up, occurs dynamically not afterwards and as such can help the organization decide what to do next instead of what happened. In a self-organizing system, feedback is gathered about those things the organization is committed to. In a higher education setting, this could be about the quality of its assessments, or collaborative process, or the extent to which graduate attributes are embedded in the design of programs, etc. Most important, the feedback emerges from the normal work of being a faculty member, leader or student.

Feedback can have major positive or negative effects on a system. Any time new rules are introduced, a system relies on feedback so that the effects may be amplified and implemented with the longer-term aim of becoming a permanent fixture (Morrison 2008; MacIntosh and MacLean 2006). Obviously feedback would be particularly critical during this time. The disequilibrium that is created through feedback processes interacts to produce unique forms of order (Houchin and MacLean 2005). Feedback among agents can alter system rules.

Human organisations are considered nonlinear feedback systems that contain certain rules and multiple relationships. Stacey (2006) suggested that there are particular laws within organisations that take the form of "decision rules and scripted relationships between people within an organization and with people across organizational boundaries" (p. 81). The example of eBay illustrates the concept of feedback and its impact perfectly. After a transaction has occurred via eBay, both the purchaser and the seller are asked to provide feedback regarding their experience. The comments made and the rating provided impact on each agent's "position" within the system. If negative feedback is provided this affects the percentage rating given to the agent, which can affect future interactions as potential purchasers may be wary about engaging in an interaction with an agent who is not seen as reliable. Conversely, positive feedback impacts well on an agent's rating and encourages future interactions.

In the SOU approach feedback is constant and enabled by technologies that engage the community at all levels in the design, enactment and engagement with learning and teaching. Feedback is part of that engagement and the tools employed to build and deliver learning and teaching. They are different from learning management systems in that the tools express the commitments and the organizational design and are built to express learning and teaching approaches known to influence student learning. Further, engaging with feedback generates a new form of analytic data that emerges from the agency of all involved in the learning and teaching process and does not involve surveillance of faculty and students as is the case with many contemporary models.

4.5   Dispersed Control

Dispersed control (Holland 1998) is about empowering those with agency in the system. Shared commitments, feedback and embedded design mean that all have a stake in the way the organization works. Those doing the actual work of the organization have pivotal agency in the system. In a higher education setting, an executive team may have a high-level view of the quality of assessment in the university, or the constructive alignment of courses. That view emerges from the normal work of developing quality assessments and aligning the key features in program design undertaken by faculty members. The high-level view is an emergent expression of the work and feedback of many who make the quality happen. SOU makes possible the multi-level view of those things to which the organization is committed and has embedded in its design. Everyone is paying attention to learning and teaching although no one is surveilled in that process.

Dispersed Control is also informed by networks and network theory and the way in which networks of teams can create professional small worlds (Barabási 2002) that pool collective intelligence, as people work together collaboratively. In SOU, intra and inter-individual capability and organizational culture are engendered by more collaborative organizational design. Culture building extends beyond the development of individuals and leaders to focus on the organization and the way its processes, methods and systems engender shared agency (Robertson 2014).

The benefits of dispersed control are that ownership and innovation are shared, collective intelligence is highly valued and the whole system is viewed as a professional body, rather than all "ownership" being directed towards the upper management or executive levels in a hierarchical structure. Control is dispersed and distributed and relies on the interaction of all agents, so effective communication systems are essential. As collective intelligence is so highly valued within this process, dispersed control appears as a natural progression of the respect for and acknowledgement of each agent's professionalism. One of the most challenging aspects associated with the principle of dispersed control is the need to build "the organisational structures, network and pathways required for genuine collaboration to occur" (Bain 2007, p. 55). In an educational context, dispersed control of an activity allows feedback at all levels and allows educators to be engaged in continual formal and informal professional communication.

In the SOU design phase, existing committees are reconstituted to conform to the dispersed control and self-similarity principles and a policy framework that reflects the university's commitments about learning and teaching. In implementation these teams or committees work with the new design to approve programs (in a self-accreditation scenario) and/or manage implementation according to the tenets of the organization's design. This occurs in self-similar ways at all levels of the organization.

4.6   Schema

The five principles identified here are designed to function interactively to create a common schema. A schema is an organizational structure that is developed as an individual interacts with their environment (Green 2010) and helps agents to work together to execute their particular roles within a system (Gell-Mann 1994). For example, most people have a schema for driving a car. A car schema allows people to deal automatically with the sameness in the process and to focus on any subtle changes that may occur within an individual driving experience such as poor weather, additional passengers or an unfamiliar route. People use a meta-perspective combined with a specific understanding of the regular features of this system to avoid cognitive overload, creating the potential for adaptation according to each particular circumstance. This schema is based on simple rules which ensure that agents within this system are clear about their roles and able to participate actively within the system.

The use of a schema addresses a common criticism of program and course design suggesting that there are often serious disjunctures within programs between theory and practice (Lynch 2012; Mueller and File 2015; Wrenn and Wrenn 2009) and the work of individual agents. A process based on these six principles reflects a shared schema where assumptions, processes and technologies reflect a common understanding and agency that frames the process in non-prescriptive ways. A schema and its associated professional language result from a design metaphor that is constantly tested, instantiated, and changed by feedback processes. A schema then is a dynamic entity that drives practice and evolves as agents engage repeatedly with a complex adaptive system.

Legitimate dispersed control relies upon a shared schema because agents are able to engage in genuine self-organizing behaviour (Gell-Mann 1994). While the initial cornerstones of that schema and its development may not be autogenetic (Kauffman 1995), the idea of the self-organizing university is that those initial design cornerstones are subject to dynamic and ongoing adaptation and evolution. The schema is subject to constant change based upon emergent feedback. The schema provides the form required for shared understandings, the creation of a genuine community of practice, and the flexibility required for individual agency within that community.

When applied to educational settings the principles represent a model for the self-organizing university (SOU). In SOU, the principles provide design guidance for creating a self-organizing university. The simple rules make possible the elusive articulation of what the organization stands for, its big ideas, commitments, and values, while embedded design drives the way those high-level commitments are articulated in the organization's design and ultimately its day-to-day work. In a higher education context, commitments drive the design of more articulated role descriptions, new policies, and the performance management methods. The emergent feedback principle also expresses commitments, beliefs and values about what the organization stands for and in the way feedback expresses agency in the organization (Bain and Drengenberg 2014). Feedback is the focus of the way the organization is governed, what happens at meetings and in the way the organization problem solves. The focus of feedback is on the things the organization values—quality learning design, collaboration, and quality learning and teaching.

The dispersed control and similarity at scale principles inform the way the organization is designed, and specifically the governance structure. This includes the self-similar roles of faculty and university committees and the way control for decision-making, informed by feedback, is dispersed broadly across the community. This includes broader stakeholder participation in the program design process. Program design is collaborative and inclusive, control is dispersed and there is a self-similar focus on the work of key groups across different levels of the organization—program team, faculty and university.

Table 4.1 summarizes the ways in which the theoretical principles are, and have been employed in the SOU.

The five theoretical principles addressed in Table 4.1, function interactively to create a common schema. Each of the examples in the table is represented as a discrete entity for ease of understanding. Equally important is the way they interact. Figure 4.1 shows the SOU Process and the way it is deployed; beginning with the development of commitments and showing the way each principle is applied to create a schema for learning and teaching.

A key feature of the SOU is the way theoretical concepts can be seen in any element of the design and each in all. For example, software tools designed and developed to implement the approach incorporate simple rules about specific research-based practices. In combination the tools and practices reflects the way groups and teams function within the academic governance process, which in turn reflects the policy and in the way faculty members are recognised and rewarded. Reference to all of these elements—governance, policy, tools can be seen in each of these elements and each in all. This self-similarity results in a dynamic and constantly evolving organization. The agency of those involved working with a shared schema generates ongoing emergent feedback that produces change in the schema and adaption.

Table 4.1 Theoretical principles in action in the SOU

Principle	Example application	Role
Simple rules	Commitments underpin learning and teaching policies, program approval policies and graduate learning outcomes	Articulate university strategy in values, set the cornerstone for the design of learning and teaching at the university level
Embedded design	Academic and leadership role descriptions; academic governance meeting process structure; software tools for program design	Commitments are embedded in all elements of the University design. Each echoes the other- all are in each (e.g., collaboration is required in the meeting process, to use the tools, instantiated in the policies, and in the role descriptions). Graduate learning outcomes are reflected in commitments, in standards, in role descriptions, in the work process created by technologies and in the feedback expectations
Similarity at scale	Role and function of committee process, collaborative process at all levels of the organization	Simplify the organizational design and focus on the commitments. The terms of reference of university committees are self-similar to those of faculty committees. Leadership position descriptions are self-similar to those of academics and divisional staff. Their roles (committees and individuals) may vary around responsibility and level of leadership, although they focus on the same self-similar things
Emergent feedback	Articulating responsibilities of stakeholder groups; feedback processes embedded in software; feedback focuses on commitments reflected in the policy	Empower the agency of those actively involved in the feedback process. Ensure feedback reflects commitments and the design. Happens as part of the normal work and a time proximal to that work. Ensure that data is aligned with normal work and emerges from that work and is not about what happened
Dispersed control	Program team collaborative decision-making, transparency of feedback processes	Empower the individuals doing the challenging work on the ground to drive the process, provide the critical feedback and drive the expression of commitments in their day-to-day work. Ensure that the governance structure functions bottom-up and focuses on commitments—simplify up (Drengenberg 2013; Latour 2010)

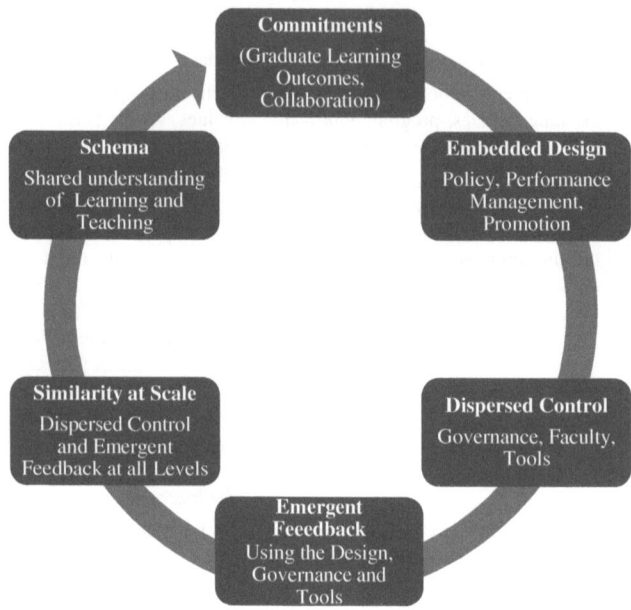

Fig. 4.1 The SOU process of connecting theoretical principles and practice

References

Allen, J. M., & Wright, S. E. (2014). Integrating theory and practice in the pre-service teacher education practicum. *Teachers and Teaching: Theory and Practice, 20*(2), 136–151. doi:10.1080/13540602.2013.848568.

Andrews, M., Pritchett, L., & Woolcock, M. (2012). Escaping capability traps through problem driven iterative adaptation. *Centre for International Development at Harvard University* (CID Working paper No.240).

Bain, A. (2007). *The self-organizing school: Next-generation comprehensive school reforms.* Lanham, MD: Rowman & Littlefield.

Bain, A., & Drengenberg, N. (2014, September). *Emergent feedback: A smart learning view.* Paper presented at the U-Imagine Learning Analytics Symposium, Wagga Wagga, NSW.

Bain, A., & Swan, G. (2011). Technology enhanced feedback tools as a knowledge management mechanism for supporting professional growth and school reform. *Educational Technology Research and Development, 59*(5), 673–685. doi:10.1007/s11423-011-9201-x.

Bain, A., & Weston, M. E. (2012). *The learning edge: What technology can do to educate all children.* New York: Teachers College Press.

Barabási, A. (2002). *Linked: The new science of networks.* New York: Perseus.

Chao, I. T., Saj, T., & Hamilton, D. (2010). Using collaborative course development to achieve online course quality standards. *The International Review of Research in Open and Distance Learning, 11*(3), 106–126.

Cherry, K. (2014). *Selected quotations by Kurt Lewin* [online newspaper article]. Retrieved from http://psychology.about.com/od/psychologyquotes/a/lewinquotes.htm.

Coate, K., & Tooher, M. (2010). The Galway symposium on design for learning: Curriculum and assessment in higher education. *Teaching in Higher Education, 15*(3), 347–354. doi:10.1080/13562511003740924.

Drengenberg, N. (2013). *Joining the dots: Simplifying up.* Unpublished manuscript, Australia University, Bathurst.

Fitzmaurice, M. (2010). Considering teaching in higher education as a practice. *Teaching in Higher Education, 15*(1), 45–55. doi:10.1080/13562510903487941.

Gell-Mann, M. (1994). *The quark and the jaguar: Adventures in the simple and the complex.* New York: W. H. Freeman and Co.

Gleick, J. (2008). *Chaos: Making a new science* (2nd ed.). New York: Penguin Books.

Green, B. A. (2010). Understand schema, understand difference. *Journal of Instructional Psychology, 37*(2), 133–145.

Holland, J. H. (1998). *Emergence: From chaos to order.* Reading, MA: Perseus.

Houchin, K., & MacLean, D. (2005). Complexity theory and strategic change: An empirically informed critique. *British Journal of Management, 16*, 149–166. doi:10.1111/j.1467-8551.2005.00427.x.

Johnson, S. (2001). *Emergence. The connected lives of ants, brains, cities, and software.* New York: Simon and Schuster.

Juarrero, A., & Rubino, C. A. (2010). *Emergence, complexity and self-organization: Precursors and prototypes (exploring complexity).* Naples, Fl: ISCE Publishing.

Kauffman, S. (1995). *At home in the universe: The search for laws of self-organization and complexity.* New York: Oxford University.

Krugman, P. (1996). *The self-organizing economy.* Oxford: Blackwell.

Kuhn, T. (1996). *The structure of scientific revolutions* (3rd ed.). Chicago: University of Chicago.

Latour, B. (2010). Tarde's idea of quantification. In M. Candea (Ed.), *The social after Gabriel Tarde: Debates and assessments* (pp. 145–162). London: Routledge.

Lester, J., & Kezar, A. J. (2012). Understanding the formation, functions, and challenges of grassroots leadership teams. *Innovative Higher Education, 37*(2), 105–124. doi:10.1007/s10755-011-9191-y.

Levine, T. H., & Marcus, A. S. (2010). How the structure and focus of teachers' collaborative activities facilitate and constrain teacher learning. *Teaching and Teacher Education, 26*, 389–398. doi:10.1016/j.tate.2009.03.001.

Levy, P. (2001). *Cyberculture.* Minneapolis: University of Minnesota Press.

Lynch, D. E. (2012). *Preparing teachers in times of change: Teaching schools, standards, new content and evidence.* Brisbane, Australia: Primrose Hall Publishing Group.

MacIntosh, R., & MacLean, D. (2006). Conditioned emergence: A dissipative structures approach to transformation. In R. MacIntosh, D. MacLean, R. Stacey, & D. Griffin (Eds.), *Complexity and Organization: Readings and conversations* (pp. 27–55). London: Routledge.

Marshall, S. J., Orrell, J., Cameron, A., Bosanquet, A., & Thomas, S. (2011). Leading and managing learning and teaching in higher education. *Higher Education Research & Development, 30*(2), 87–103. doi:10.1080/07294360.2010.512631.

Merry, U. (1995). *Coping with uncertainty. Insights from the new sciences of chaos, self-organization, and Complexity.* Westport, CT: Praeger.

Morrison, K. (2002). *School leadership and complexity theory.* London: Routledge Falmer.

Morrison, K. (2008). Educational philosophy and the challenge of complexity theory. In M. Mason (Ed.), *Complexity theory and the philosophy of education* (pp. 16–31). West Sussex, UK: Wiley-Blackwell.

Mueller, J. J., & File, N. K. (2015). Teacher preparation in changing times: One program's journey toward re-vision and revision. *Journal of Early Childhood Teacher Education, 36*(2), 175–192. doi:10.1080/10901027.2015.1030521.

Pascale, R. T., Millemann, M., & Gioja, L. (2000). *The new laws of nature and the new laws of business.* New York: Crown Publishing Group.

Perry, J. A. (2014). Changing schools of education through grassroots faculty-led change. *Innovative Higher Education, 39*(2), 155–168. doi:10.1007/s10755-013-9267-y.

Prigogine, I., & Stengers, I. (1984). *Order out of chaos: Man's new dialogue with nature.* New York: Bantam Books.

Prokopenko, M. (2013). Design vs self-organization. In M. Prokopenko (Ed.), *Advances in applied self-organizing systems* (2nd ed., pp. 3–17). London, UK: Springer.

Robertson, B. (2014). *Differentiating organization and tribe*. [Web log post]. Retrieved from http://holacracy.org/blog/differentiating-organization-tribe.

Scott, G. P. (1991). *Time rhythm and chaos: In the new dialogue with nature*. Ames, Iowa: Iowa State University Press.

Seel, R. (2000). Culture and complexity: New insights into organizational change. *Organization & People, 7*(2), 2–9.

Stacey, R. (2006). The science of complexity: An alternative perspective for strategic change processes. In R. MacIntosh, D. MacLean, R. Stacey, & D. Griffin (Eds.), *Complexity and organization: Readings and conversations* (pp. 74–100). London: Routledge.

Sull, D., & Eisenhardt, K. (2012). Simple rules for a complex world. *Harvard Business Review,* 69–74.

Tom, A. R. (1997). *Redesigning teacher education*. Albany, NY: State University of New York Press.

Waldrop, M. M. (1993). *Complexity: The emerging science at the edge of order and chaos*. New York: Touchstone, Simon and Schuster.

Wrenn, J., & Wrenn, B. (2009). Enhancing learning by integrating theory and practice. *International Journal of Teaching and Learning in Higher Education, 21*(2), 258–265.

Zundans-Fraser, L., & Bain, A. (2015a). The role of collaboration in a comprehensive programme design process in inclusive education. *International Journal of Inclusive Education.* Advance online publication. doi:10.1080/13603116.2015.1075610.

Zundans-Fraser, L., & Bain, A. (2015b). *How do institutional practices for program design and review address areas of need in higher education.* Manuscript submitted for publication.

Zundans-Fraser, L., & Lancaster, J. (2012). Enhancing the inclusive self-efficacy of pre-service teachers through embedded course design. *Education Research International.* doi:10.1155/2012/581352.

Chapter 5
Research

Abstract This chapter takes up the specific approaches to learning and teaching that underpin the self-organizing university. The role of these curricular and pedagogical innovations is described along with the research that supports their inclusion. The chapter is designed to show how the model is informed by practical well-researched approaches to learning and teaching.

Keywords Learning intention · Mastery · Collaboration · Teaching approaches · Constructive alignment · Design theory · Feedback · Assessment

The six theoretical design principles in combination with new technologies provide a framework for designing an organization for learning and teaching. However, the design theory does not direct an institution regarding the *content* to which the framework is applied and the vision and practice of learning and teaching that will drive its simple rules and instantiate its embedded design. Critical to SOU is answering the "what" questions: What are the simple rules about? What is the university committed to in terms of educational practice? What learning and teaching approaches are included in software tools? What are the processes, approaches and practices that will be embedded in policies, role descriptions, promotion approach, etc.? A key additional question pertains to what the organization believes and values about learning and teaching and how it sources that information?

The SOU employs over sixty years of modelling and research related to factors that influence student learning to answer the "what" questions. As such, the content of SOU has a broad and deep base in longitudinal international research about learning and teaching that defines the content of the model and interacts with the theoretical principles. Most important, are the connections between theory and research whereby the theory connects with research to drive the design of the SOU approach.

It is important to emphasise that the SOU enables a university to answer its "what" questions in its own way. While a strong case can be made for the research approaches described in the following section, the model does not prescribe a set of practices. Each institution should build its own case for approaches that suit its needs and context. The SOU does suggest that policies, role descriptions and governance models should be informed by approaches that have been subject to

A. Bain and L. Zundans-Fraser, *Rising to the Challenge of Transforming Higher Education*, SpringerBriefs in Education, DOI 10.1007/978-981-10-0261-8_5

longitudinal empirical scrutiny and have an applied track record of improving the quality of learning and teaching. The practices described in the examples that follow meet these criteria. The model employed to undertake the integration of learning and teaching research in the SOU is also described.

5.1   Models and Analyses of Learning and Teaching

This title refers broadly to work that has sought to identify at a high level those factors that contribute to student learning. It includes the work of Carroll (1963) who focused on the role of perseverance, instructional quality, aptitude and opportunity in the 1960's. Bloom (1976) examined factors including, instructional quality, affective factors and aptitude. Walberg (1986) examined the role and dimensions of teaching that influences student learning. This also included the role of aptitude, quality of instruction, motivation and environmental factors. More recently, Hattie (2009) undertook a synthesis of over 800 meta-analyses relating to student achievement, identifying those factors that exert more or less of an influence on student learning.

The models and analyses described here and others (e.g., Brophy and Good 1986; Marzano 1998; Scheerens 1992) reflect conclusions drawn from literally thousands of discrete research efforts that identify those factors that influence learning and provide an initial focus for the content of a design. In the SOU, these models and analyses serve to map the content territory and identify foci that inform the selection of specific learning and teaching approaches, the identification of feedback sources and methods and the approach to analytics, the design of software, and the creation of professional development activity and resources. They provide a content focus for the design principles described in the previous section. This includes informing the identification of commitments, the approaches that are embedded in the design of the organization and in the focus of feedback. Further, those bodies of literature address variables that interact in their contribution to the content of the SOU approach. For example, feedback and assessment are made more or less meaningful depending upon the learning context in which they are placed. Constructive alignment (Biggs 1996) in courseware design is dependent upon the quality of the elements being aligned. The SOU, as a transformational organizational design approach, is a way to create the supportive learning and teaching context required to ensure that the discrete features of feedback or successful teaching practice are likely to be maximised. The models of learning and teaching are instantiated by specific practices and strategies that provide the content for the SOU. The sections that follow describe examples of those approaches in the areas of goal setting, constructive alignment, feedback and assessment, and teaching approaches.

5.2 Learning Intention and Process

Hattie (2009) describes a range of strategies that improve learning by focusing clearly on learning intention and process. Those approaches include goal setting, mapping and organisers, learning hierarchies, mastery learning, and worked examples. The central idea of this research focuses on the power associated with making clear the goals, process and expected outcomes of learning. These approaches cover a broad range of cognitive and meta-cognitive learning strategies including advance organization, concept mapping, self-questioning, self-monitoring, and sequencing modeling and practice in worked examples (de Boer et al. 2013; Dignath et al. 2007; Hattie 2009; Hattie et al. 1996). These meta-analyses and others incorporate research from thousands of studies that have investigated the discrete elements of learner intention showing that they make a profound and powerful contribution to student learning. They have produced effect sizes in comparative studies of 0.41 for advance organisers, 0.57 for concept mapping, 0.58 for mastery learning and 0.56 for goal setting (Hattie 2009). At the core of the connections made among these different areas of research, is the straightforward idea that articulating intent and learning process creates conditions that are more conducive to successful learning by clarifying expectations and making visible what learners are asked to do. This is especially the case for learners who do not have the background and related meta-cognitive skills required to deconstruct and interpret less intentional learning experiences.

In the SOU, this research on learning intention and process is embedded at all levels. For example, policy is framed using this research, role descriptions reflect the policy, technology tools feature the research drivers for learning success. Further, the organizational change dimensions of the SOU, including the approach to governance, policy, promotion, performance management are designed to create the context in which the potential of those practices can be amplified. This is done by making an extant connection to the research in role descriptions, performance management expectations and the institutional framework that governs approval processes. While Hattie (2009) and Marzano (1998) described the immediate learning and teaching conditions required for successful practice, the SOU takes up the organizational context required for scalable application in a university setting. This includes providing the broader setting conditions that support, incentivize and empower a community to engage with more sophisticated practice.

5.3 Constructive Alignment

Constructive alignment (Biggs 1996) can be seen as an extension of the body of work related to learning intention described by Hattie (2009). The concept involves building extant connections between learning intention or expected outcomes, teaching activity/learning experience and assessment in the development of learning experiences. Constructive alignment has a common sense appeal and has garnered significant support in higher education (Biggs and Tang 2007; Larkin and Richardson 2013; Teater 2011). The empirical support for this approach is reflected in its connection to the clarification of learner intent and in additional work that has examined its efficacy as a discrete practice (Kenney 2012; Treleaven and Voola 2008). While not as extensive as other areas identified throughout this review (e.g., cooperation or feedback), there is strong support for designing learning and curricular experiences employing this principle. A number of studies have shown improved satisfaction and learning outcomes for courses and programs that applied the principle of constructive alignment (Kuhn and Rundle-Thiele 2009; Vanfretti and Farrokhabadi 2013; Wang et al. 2013).

In the SOU, constructive alignment is a pivotal feature of the learning design process. Constructive alignment can be seen as a mechanism for enacting the theory in practice—a commitment to constructive alignment is embedded in the tools, policy, and feedback in ways that are self-similar across program and course levels. This research is also a key focus of the governance process where constructive alignment when embedded in learning and teaching policy can become an important term of reference for teams or groups approving programs and courses or monitoring their implementation.

5.4   Feedback and Assessment

Feedback is a critical factor in predicting student learning. The research on feedback is extensive and has identified the influence of form, timing, type, and source as determinants of feedback effectiveness (Brinko 1993; Scheeler et al. 2004). Feedback is not a standalone contributor to student learning. Its effectiveness is related to a range of contextual factors related to the broader circumstances under which feedback is shared (Hattie 2009; Marzano 1998). The effectiveness of feedback is contingent upon the overall quality of the learning experience—the quality of the assessment task and the compendium of influences related to learning intention and process (Hattie 2009). Feedback has produced effect sizes ranging from 0.73 (Hattie 2009) to 1.13 (Marzano 1998).

The SOU takes up the role of feedback in two ways. First, as described earlier, feedback is placed in a meaningful context as a key component of a self-organizing system; feedback is secured within the broader organizational design context where commitments are clear, where embedded design enacts those commitments, where control is dispersed and where feedback can operate in emergent self-similar ways. Second, and as a consequence of the design, feedback is deeply embedded in the work of teams at all levels—those working to develop programs and courses and those at the university level approving programs. Feedback emerges from building commitments, integrating standards, developing assessment tasks, courses and modules. Program team members share feedback about the way the team has gone about completing these design tasks. In all cases, feedback is an emergent expression of the normal work of individuals and groups. The research on when and how feedback is effective informs the way emergent feedback is structured in the SOU. Feedback in the SOU comes from multiple sources, focuses on the role of peers and subordinates, encourages immediacy, is evidence-based and positively focused (Brinko 1993; Hattie 2009; Marzano 1998; Scheeler et al. 2004).

In the SOU, feedback also operates in concert with the other research-based practices and approaches described here in the form of criterion and standards based assessment (CSBA). Criterion and standards based assessment includes a number of the learning intention features described earlier including goal-setting, mapping and scaffolding (Hattie 2009). Research on, and analysis of, CSBA indicates that its success is contingent upon contextual factors including the provision of exemplars and its integration into the broader instructional milieu (Donovan et al. 2001; Sadler 2005). The SOU creates the broader contextual influences necessary for making CSBA effective.

Unique to the SOU is the way in which technology, when applied to the normal day-to-day work of a self-organizing system, can make feedback emerge from that routine activity and the normal work of the organization. A study in progress is examining the effects of feedback on teaching at scale in environments using the Edge Technology Tools (described subsequently) (Bain and Weston 2015). In the SOU, feedback is less about what happened and more about what is happening (Bain 2007).

5.5   Teaching Approaches

One of the key findings emerging from the meta-analyses of factors that influence student learning is that there is immense variability in the efficacy of different teaching approaches. For example, lectures and mastery teaching are both forms of teacher led instruction. Lectures have been shown to exert only a limited effect on achievement. Mastery learning (Hunter 2004) is a powerful source of achievement effects. Similarly, cooperative learning that pays attention to the structure of the experience, individual accountability and interdependence among participants is a strong predictor of achievement (Hattie 2009; Johnson et al. 2014; Marzano 1998; Slavin 1996). Group work much less so. Other teaching approaches that have strong student learning effects include peer mediation, reciprocal teaching, and cognitive and meta-cognitive strategies, especially when they are taught explicitly (Rosenshine 1997).

The SOU emphasises the identification of those approaches that have strong effects on student learning to feature in the model. Technology tools help to build experiences that employ those approaches. Feedback on program design can prioritize these approaches while policies can emphasize practices that have a track record of success. Doing so ensures that those features are represented in both the design and delivery of learning activities. Table 5.1 summarizes the foundational

Table 5.1 SOU foundational research

Principle	Description	Role in the SOU	Research
Models of learning	Refers to researchers who have modelled the factors that contribute to student learning	These researchers mapped the territory related to factors influencing learning. They provide a broad-brush framework for the foci of the SOU especially related to the development of simple rules, analytics framework, and to drive the embedded design process	Bloom (1976); Brophy and Good (1986); Carroll (1963); Hattie (2009); Scheerens (1992); Walberg (1986)
Goal setting frameworks, scaffolding and mapping	Research on creating an extant process for goal setting, scaffolding and framing at all levels of the organization and in all activities	This work provides the basis for the design of policy, feedback and the context for learning design in the SOU	de Boer et al. (2013); Dignath et al. (2007); Hattie (2009); Hattie et al. (1996)
Constructive alignment	A widely accepted design principle to make learning more accessible	Applied in the SOU to program and course design	Biggs and Tang (2007, 2011); Kenney (2012); Larkin and Richardson (2013); Teater (2011); Treleaven and Voola (2008).
Feedback and assessment	Feedback effectiveness relates to a range of contextual factors and is contingent on the quality of the assessment tasks, learning intentions and process	Feedback is used in the broader organizational design in a self-similar manner, is embedded in the work of teams at all levels within an organization and within assessment processes	Brinko (1993); Hattie (2009); Hattie and Timperley (2007); Marzano (1998); Sadler (2005); Scheller et al. (2004)
Teaching approaches	These are approaches that have been shown to improve student learning. This is an ongoing process employed to seek out those approaches that when reconciled with content and technology can leverage the university's learning and teaching value proposition	Embedded in the design of software tools, including feedback processes, policy and implementation	Hattie (2009); Johnson et al. (1981); Kyndt et al. (2013); Rosenshine (1997, 2012); Slavin (1996); Springer et al. (1999)

research utilised to inform the SOU approach. The research described here constitutes a sample of the extensive work in these areas.

The practices described here are used to flesh out the learning and teaching design by instantiating the theoretical framework with known efficacious learning and teaching approaches. The interaction of the theoretical principles and research creates the conditions for the design of an adaptive system. The tools, the policies, the governance process are each and all designed to situate successful practice in an organizational design where those practices can interact and contribute successfully to student learning. For example, a collaborative organizational process where meetings function successfully to pool the capacity of participants, or tools that make feedback about effective practice possible, create a context for solving problems about the use of a teaching approach. Collaborative process and research-based pedagogy interact. As was the case with the theory, the practices in the SOU are designed to interact with each other because the organizational design makes that interaction possible. An adaptive design may employ different bodies of research, employ different empirical traditions that produce different approaches and emphases in a theory-research driven approach to change. The main points for consideration here are twofold. First, that serious thought is given to the antecedents and underpinnings of the overall approach irrespective of the traditions and research identified as being relevant and of greatest importance. Second, that the dots get joined when applying research. Embedded design is employed to situate that research, in all facets of the organization.

References

Bain, A. (2007). *The self-organizing school: Next-generation comprehensive school reforms.* Lanham, MD: Rowman & Littlefield.

Bain, A., & Weston, M. (2015). *The scalable effects of feedback on the quality of teaching: A technology mediated approach.* Manuscript in preparation.

Biggs, J. (1996). Enhancing teaching through constructive alignment. *Higher Education, 32*(3), 347–364.

Biggs, J., & Tang, C. (2007). *Teaching for quality learning at university.* Maidenhead, UK: McGraw-Hill and Open University Press.

Biggs, J., & Tang, C. (2011). *Teaching for quality learning at university* (4th ed.). Maidenhead, UK: Open University Press.

Bloom, B. S. (1976). *Human characteristics and school learning.* New York: McGraw-Hill.

Brinko, K. (1993). The practice of giving feedback to improve teaching. What is effective? *Journal of Higher Education, 64*(5), 574–593.

Brophy, J., & Good, T. L. (1986). Teacher behavior and student achievement. In M. C. Wittrock (Ed.), *Handbook of research on teaching* (pp. 328–375). New York: Macmillan.

Carroll, J. B. (1963). A model of school learning. *Teachers College Record, 64,* 723–733.

De Boer, H., Donker-Bergstra, A. S., & Kostons, D. N. M. (2013). *Effective strategies for self-regulated learning: A meta-analysis.* Retrieved from Gronings Instituut voor Onderzoek van Onderwijs: http://gion.gmw.eldoc.ub.rug.nl/FILES/root/2013/EffectiveStrategies/Effective Strategies.pdf.

Dignath, C., Büttner, G., & Langfeldt, H. P. (2007). *The efficacy of self-regulated learning interventions at primary and secondary school level: A meta-analysis.* Budapest, Hungary: Paper presented at the European Association on Learning and Instruction.

Donovan, B., Price, M., & Rust, C. (2001). The student experience of criterion-referenced assessment (through the introduction of a common criteria assessment grid). *Innovations in Education and Teaching International, 38*(1), 74–85. doi:10.1080/147032901300002873.

Hattie, J. (2009). *Visible learning: A synthesis of over 800 meta-analyses relating to achievement.* New York: Routledge.

Hattie, J., Biggs, J., & Purdie, N. (1996). Effects of learning skills interventions on student learning: A meta-analysis. *Review of Educational Research, 66*(2), 99–136. doi:10.3102/00346543066002099.

Hattie, J., & Timperley, H. (2007). The power of feedback. *Review of Educational research, 77* (81), 81–112. doi:10.3102/003465430298487.

Hunter, R. (2004). *Madeline Hunter's mastery teaching: Increasing instructional effectiveness in elementary and secondary schools.* Thousand Oaks, CA: Corwin.

Johnson, D. W., Johnson, R. T., & Smith, K. A. (2014). Cooperative learning: Improving university instruction by basing practice on validated theory. *Journal on Excellence in College Teaching, 25*(3 & 4), 85–118.

Johnson, D., Maruyama, G., Johnson, R., Nelson, D., & Skon, L. (1981). Effects of cooperative, competitive and individualistic goal structures on achievement: A meta-analysis. *Psychological Bulletin, 89*(1), 47–62.

Kenney, J. L. (2012). Getting results: Small changes, big cohorts and technology. *Higher Education Research & Development, 31*(6), 873–889. doi:10.1080/07294360.2012.672402.

Kuhn, K. A. L., & Rundle-Thiele, S. R. (2009). Curriculum alignment: Exploring student perception of learning achievement measures. *International Journal of Teaching and Learning in Higher Education, 21*(3), 351–361.

Kyndt, E., Raes, E., Lismont, B., Timmers, F., Cascallar, E., & Dochy, D. (2013). A meta-analysis of the effects of face-to-face cooperative learning: Do recent studies falsify or verify earlier findings? *Educational Research Review, 10.* doi:10.1016/j.edurev.2013.02.002.

Larkin, H., & Richardson, B. (2013). Creating high challenge/high support academic environments through constructive alignment: Student outcomes. *Teaching in Higher Education, 18*(2), 192–204. doi:10.1080/13562517.2012.696541.

Marzano, R. (1998). *A theory-based meta-analysis of research on instruction.* Aurora, CO: Mid-continent Research for Education and Learning. Retrieved from http://www.peecworks.org/peec/peec_research/I01795EFA.2/Marzano%20Instruction%20Meta_An.pdf.

Rosenshine, B. (1997). *The case for explicit, teacher-led, cognitive strategy instruction.* Paper presented at the annual meeting of the American Educational Research Association, Chicago, IL.

Rosenshine, B. (2012). Principles of instruction: Research-based strategies that all teachers should know. *American Educator, 36*(1), 12–19.

Sadler, R. (2005). Interpretations of criteria-based assessment and grading in higher education. *Assessment & Evaluation in Higher Education, 30*(2), 175–194. doi:10.1080/0260293042000264262.

Scheeler, M., Ruhl, K., & McAfee, J. (2004). Providing performance feedback to teachers: A review. *Teacher Education and Special Education, 27*(4), 396–407.

Scheerens, J. (1992). *Effective schooling: Research, theory and practice.* New York: Cassell.

Slavin, R. E. (1996). Research on cooperative learning and achievement: What we know, what we need to know. *Contemporary Educational Psychology, 21*(1), 43–69.

Springer, L., Stanne, M. E., & Donovan, S. S. (1999). Effects of small-group learning on undergraduates in science, mathematics, engineering and technology: A meta-analysis. *Review of Educational Research, 69*(1), 21–51. doi:10.3102/00346543069001021.

Teater, B. (2011). Maximizing student learning: A case example of applying teaching and learning theory in social work education. *Social Work Education, 30*(5), 571–585.

Treleaven, L., & Voola, E. (2008). Integrating the development of graduate attributes through constructive alignment. *Journal of Marketing Education, 30*(2), 160–173. doi:10.1177/0273475308319352.

Vanfretti, L., & Farrokhabadi, M. (2013). Evaluating constructive alignment theory implementation in a power systems analysis course through repertory grids. *IEEE Transactions on Education, 56*(4), 443–452. doi:10.1109/TE.2013.2255876.

Walberg, H. J. (1986). Synthesis of research on teaching. In M. C. Wittrock (Ed.), *Handbook of research on teaching*. New York: Macmillan.

Wang, X., Su, Y., Cheung, S., Wong, E., & Kwong, T. (2013). An exploration of Biggs' constructive alignment in course design and its impact on students' learning approaches. *Assessment & Evaluation in Higher Education, 38*(4), 477–491. doi:10.1080/02602938.2012.658018.

Chapter 6
Design

Abstract This chapter describes the design process including examples of what the self-organizing university looks like in practice in the areas of policy, role descriptions, performance management, and governance. The examples include an account of the connections with theory and research. The critical role of technology is described including examples of tools for program and course design. A new way of thinking about the role of technology is introduced that includes an approach to learning analytics focused on the real day-to-day work of learning and teaching.

Keywords Policy · Performance management · Teams · Governance · Roles · Program design · Edge technology · Learning analytics

So far, we have provided discrete exemplars of the way theory and research informed a design framework using specific design decisions in SOU as examples. A remaining question is: What do the key parts of the redesigned organization look like as a whole when governance, policy, technology, roles, performance management and career trajectory become part of a complete organizational design? In an effort to build a more complete picture, we have employed excerpts from design elements in the form of policy, role descriptions, career path, team responsibilities and technologies to illustrate the product outcomes of using the SOU approach. At the outset we acknowledge that specific nomenclature for roles and working groups vary from institution to institution and readers need not stipulate around the terminology used in the excerpts. The most important takeaway is the scope of the design effort, the connections made, and what the design elements might look like in practice. Figure 6.1 describes the scope of the SOU design in four broad categories. This section will unpack those elements.

6.1 Policy

A learning and teaching policy which may already exist in some form in the university is the term of reference for the SOU design. It is where the simple rules or commitments are expressed in sufficient detail to drive the design of the

© The Author(s) 2016

A. Bain and L. Zundans-Fraser, *Rising to the Challenge of Transforming Higher Education*, SpringerBriefs in Education,
DOI 10.1007/978-981-10-0261-8_6

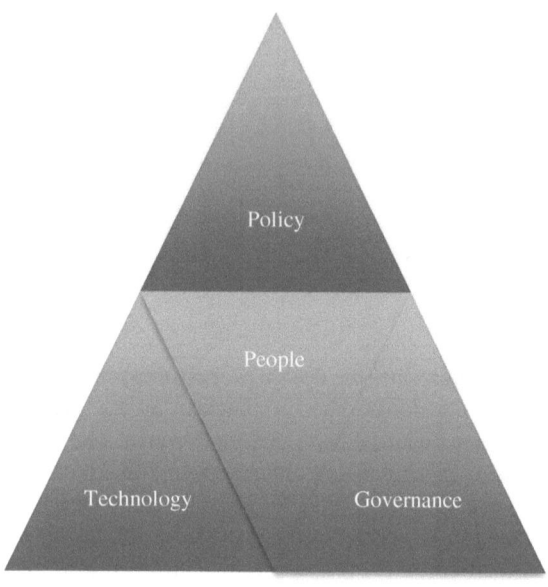

Fig. 6.1 Scope of the SOU in practice

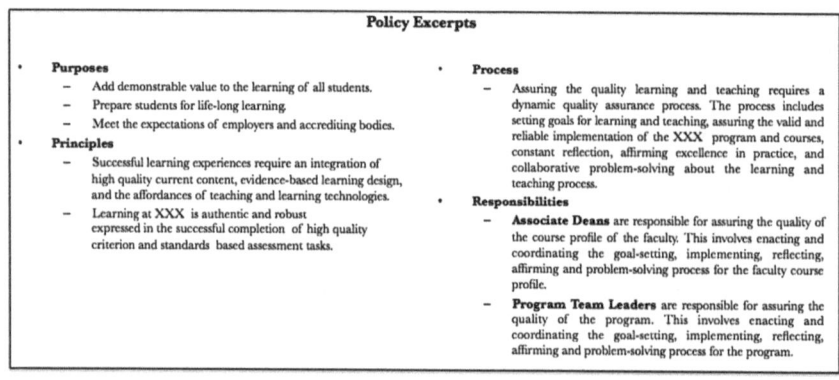

Fig. 6.2 Excerpt from the SOU policy

organization. A SOU policy may be more specific in its reference to learning and teaching practice and includes features that are readily translated into position descriptions, the role of governance groups, and technologies. Figure 6.2 shows excerpts from a SOU designed learning and teaching policy that includes sections on purposes, principles, processes and responsibilities.

The four categories in the SOU policy create a driver for embedding the policy in all elements of the model. For example, the purpose statement that describes "*adding demonstrable value to the student experience*" signals that the university intends to show how it adds value over aptitude or student entry characteristics. To do so will require a powerful approach to analytics and feedback. The principles section allows for specific reference to the learning and teaching research including criterion and standards based assessment. This signals that processes for designing and enacting programs and courses will include reference to quality criterion-based assessment tasks. The process dimension of the policy frames out a commitment to a specific type of governance activity by stating how quality assurance will happen. The responsibilities dimension creates an easy segue to the way role descriptions will be defined in the SOU design by stating what leaders in this instance are expected to do.

What is in evidence here is the application of the embedded design and the self-similarity principles. Policy is developed in a manner that makes possible the translation of intentions into action through substantive processes, methods and tools. Collectively, they reflect the "*each in all and all in each*" aspect of the SOU design model. As such, a SOU policy may be similar to existing policy documents. The categories of purpose, principle, process and responsibility are not uncommon in policy development frameworks. More important is the way those categories reflect specific commitments, are research-informed, and drive a design process.

6.2 People

The people dimension of the SOU design refers to the way the model is reflected in role descriptions, performance management and career paths to support faculty members as they migrate to new ways of addressing learning and teaching. The design addresses the critical need, described in the literature and earlier in the brief, to align institutional process and practice with a more responsive and contemporary vision of what learning and teaching means. This section illustrates the embedded design principle by showing how policy is embedded in the human resource dimensions of the organization. Figure 6.3 describes an excerpt from a role description for a faculty member leading and coordinating a degree program. The person in the role is responsible for the design, enactment and overall performance of the program.

The first point of emphasis about this excerpt is the way the description of the implementation process is an echo of the policy exemplar. The initial bullet point is a mirror of the description of the quality assurance process described in Fig. 6.2. As well as echoing the policy, the position description goes further by explaining in greater detail what that process means by defining what the role involves for sharing feedback and the way student performance is monitored. The description also refers to the technology tools used in the process. These tools will be described later in the brief.

Another key feature relates to self-similarity across roles. If we replaced the term *Program* with *Faculty Course Profile* and *Program Team Members* with *Program Directors,* this excerpt from the Program Director's role would be the same for an

---

Excerpt from Program Director Role

For Program Implementation:
    (i) Lead and Manage the Program Implementation Process

- the implementation of the quality assurance process and the performance of the programs they direct. This involves enacting the process of goal-setting, implementing, reflecting, affirming and problem-solving at the program level. It includes:

- Providing relevant program level feedback on learning and teaching using the TAES tools.
- Conducting regular teaching team meetings that engage the program team with quality assurance process including the performance of students and assessment tasks.
- Providing feedback at the subject level on the implementation of learning and teaching experiences.
- Providing feedback to faculty members about program and subject implementation using the TAES workspace tools.
- Conducting the Performance Management Process for Program Team members
- Sharing feedback related to performance management and career progression.
- Monitoring Program Performance in the TAES tools.
- Ensuring that action plans resulting from the Quality Assurance process are implemented.
- Sharing action plans with the Associate Dean
- Reporting regularly at Faculty meetings about the performance of the Programs they direct using TAES.
- Providing feedback to individual faculty members about overall performance

---

Fig. 6.3 Excerpt from a faculty member role description (program director)

Associate Dean's position description or the person managing learning and teaching at the faculty level. Replace the faculty level terms with University level descriptors and we have a description of the role of a Vice President (Academic) or equivalent position. The responsibilities for managing *a* program are similar to those for managing *all* the programs in a faculty or University. The role descriptions are self-similar at scale.

The "each in all, all in each" principle is also evident in the excerpts as policy is embedded in role—role mirrors policy. Further, both policy and role foreshadow the way governance works, by describing the role of an individual in the process of quality assurance—what happens in implementing a program. The way groups function collaboratively across the organization to manage the work of quality assurance can be derived from the role descriptions and policy.

Performance management in the SOU design is a no surprises reflection of the role and the overall design. When policy and roles are informed by well-established research much of the ambiguity around performance management is diminished. Goals are set in relation to collaborative practice, the quality of learning and teaching, assessment all of which have a qualitative and quantitative empirical term of reference because of the way research has informed the design.

Judgements about the way a leader conducts meetings or a team member designs a cooperative activity have a substantive evidence-based term of reference in the SOU. Generating the feedback that captures these qualities is essential and a function of the way teams work together and the role of technology in the process.

Excerpt from Career Progression: Course Design

Assistant Professor	**Associate Professor**
Collaborative Program Design:	Collaborative Program Design:
Feedback in TAES showing record of high quality feedback to peers, and supportive problem-solving for quality assurance at team meetings. Responds to feedback in a timely and thoughtful manner. Makes a significant contribution to course design as reflected in peer feedback in TAES.	Feedback in TAES shows record of advanced problerm-solving at the program and course level and leadership support of peers. Responds to feedback in a timely and thoughtful manner. Makes a significant contribution to program and course design as reflected in peer feedba ck in TAES. Evidence of leadership of learning and teaching capable of exerting faculty-wide influence.

Fig. 6.4 Excerpt from learning and teaching career progression document

With an empirically robust term of reference for giving and using feedback, career progression in relation to learning and teaching becomes a much less mysterious process. Figure 6.4 describes an excerpt from a learning and teaching progression for movement from Assistant to Associate Professor.

Most important is the way the SOU requires demonstrable knowledge and skill with the university's learning and teaching. Irrespective of levels, roles and titles which vary from institution to institution (e.g., the term program director may not be in common use), all members of the community need to build capacity with the processes, practices and methods of learning and teaching that underpin the organizational design. Administration is reconciled with agency and learning and teaching at all levels. The implications associated with the need for learning and teaching expertise are taken up in the implementation and engagement sections of the brief.

Each of the requirements described in Fig. 6.4 have clear line of sight back to the policy and the University's commitments. The record of feedback and problem solving means there is a real-time empirical record to underpin the case for promotion. As with policy and role, the progression excerpt echoes those components.

The SOU approach can address concerns about the way in which career progression in learning and teaching frequently involves one dimensional measurement (student evaluations) or ex-post-facto case and portfolio building that lacks rigorous predetermined standards and expectations of performance Baume and Yorke (2010). The SOU model makes expectations clear and transparent. When enabled by technological tools that produce a real-time record of work and feedback, the case or evidence base for promotion has an understood foundation that makes promotion much more criterion-based and legitimate as a result. Of importance is the expectation that no SOU design or process is static. For example, the second criterion for the Associate Professor level calls for a capacity to evolve the model to contribute to policy and innovation.

6.2.1 The Project Team

The work of implementation including the creation of the various implementation plans (e.g., change, personnel, evaluation, technology), design of professional development and software development is undertaken by a project team that assumes initial responsibility for the baseline assessment, design and pilot work. The team serves in an architectural and design capacity assuming the responsibility for the additional work required to undertake change within a working organization while also ensuring that the university's leadership assumes active responsibility for the initiative. The goal is to make the work of change part of the normal work of the organization as quickly as is possible. The expertise required when constituting the team includes software design for learning and teaching, professional development design and delivery, consultant expertise in program design and development, technology systems integration, and change management and communication. Leadership expertise that covers across theory, research, design and practice of large-scale change is also critical. In our work we find that a key role of a project team is interlocution connecting the parts of an organization responsible for the broad sweep of work related to parts of the SOU model. This is the work of migration—migrating from technologies that document to those that design, from generic to more specific and empirically informed human resource structures, from private practice responsibilities to collaborative teams. This requires a consultant capability within the project team to fill gaps in existing capability that ultimately build the sustainable capacity of the entities and individuals with whom they are consulting.

6.3 Governance

Governance in the SOU model is expressed in a series of cascading self-similar teams at different levels of the organization. Figure 6.5 shows the three levels at which the team function.

The teams are networked, each depending on the other to do their normal work. That dependence pivots on the work of program teams that build and implement the course. They generate the feedback on design, enactment and student performance that is employed subsequently at faculty and university level to establish University

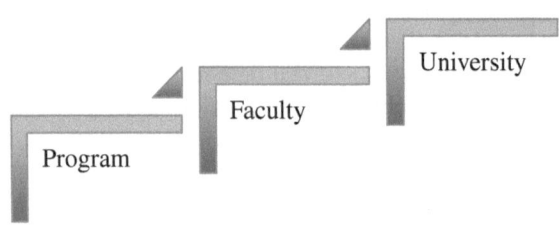

Fig. 6.5 Levels of governance in the SOU model

> **Excerpt from University level Learning and Teaching Team**
> **Terms of Reference**
>
> • **The University Learning and Teaching Team** is responsible for assuring the quality of the
> course profile of the University. This involves enacting and coordinating the goal-setting,
> implementing, reflecting, affirming and problem-solving process for the University course profile. It
> includes:
>> - Conducting regular meetings that engage the quality assurance process.
>> - Receiving regular faculty reports from Associate/Sub Deans using the TAES
>> - Monitoring Learning and Teaching Performance in TAES
>> - Reporting to Senate, The University Council, on the performance of
>>   Learning and Teaching at the University
>> - Providing feedback to Associate Deans about faculty level implementation.
>> - Assisting Associate Deans with faculty-level problem-solving
>> - Ensuring that University level action plans resulting from the Quality Assurance process
>>   are implemented.

Fig. 6.6 Excerpt from the university level learning and teaching team document indicating their responsibilities

performance. In the SOU, it is the emergent sophisticated work at ground level undertaken in the interaction among faculty members and students that empowers the system and is represented in simpler forms at higher levels within the university.

The focus of the governance model is to assure that the policy is being implemented. Figure 6.6 describes the terms of reference for a University Level Team responsible for learning and teaching.

The opening excerpt describes the connection to the process dimension of policy and specifically the quality assurance process. Just as the role description instantiated what that means for an individual, the example and specifically the sub points show the way teams work to implement the policy. At this level, the data that scales up from every program and faculty is examined at an institutional level and used as an empirical term of reference for problem solving and decision-making.

6.4 Technology

The frequent references to technology in previous sections highlight the critical role technology plays in the SOU design. In this section we explain the underpinnings of the technologies and share examples of how they work to catalyze the SOU model into action in a University setting.

Many of the technologies used by Universities (e.g., Learning and Document Management Systems) reflect and echo their prevailing organizational design, process and priorities. By this we mean the lack of focus on learning and teaching research and theory-driven organizational design are reflected in an emphasis on documentation over research-based design with tools that tend to automate an instrumental approach to the management of learning and teaching.

6.4.1   Edge Technology

Edge Technology (Bain and Weston 2012) is a model for developing technologies that improve the quality of learning and teaching and align with the SOU approach described here. To be an Edge Technology, an innovation must extend the capacity of a teacher or learner by extending the human-machine relationship in ways that complement their learning and teaching abilities. This can take the form of a teacher designing and differentiating learning experiences for multiple groups of students with differentiated content and pedagogy in ways that reduces cognitive load, maintains cohesion and a manageable learning environment. Edge Technologies also heighten collaboration and flatten communication hierarchies in educational contexts by creating the conditions for more effective collaboration and more efficient exchange of learning and teaching information (Bain and Weston 2012). They shorten the cognitive distance between those involved in the joint educational enterprise of designing and delivering learning experiences. This can involve a team of university educators integrating content, pedagogy and technology in a real-time collaboration that includes collaborative design and emergent feedback. In this way Edge Technologies can be seen as vehicles to catalyse dispersed control in a self-organizing system, as they make possible a shortening of cognitive distance to create a collaborative *small worlds* focus (Barabási 2002). If an Edge Technology can extend cognition, in doing so it should also build capacity. Users in educational contexts should learn more about design and the integration of pedagogy, content and the creation of learning experiences as a result of common use. Edge Technologies provide a technological representation of the schema that emerges from the application of the SOU model.

The model for creating Edge Technologies is described by Bain and Swan (2011, 2012), and includes using research to map solutions, component design and integration. Edge Technology emerged from a series of studies and descriptive reports undertaken over the last 14 years that examine the tools built using the Edge approach and the behavior of users and the effects of that use (see Table 9.1). They include work about enabling classroom participation, the examination of gender effects, curriculum design and use, and student and teacher effects. A major focus of this research was a large scale evaluation of the model in the Outcomes Project, a three million dollar research and design initiative that studied the uptake of the tools and their impact on teaching across three school districts in two US states (Weston and Bain 2014).

6.4.2   Tools for Adaptive Education Systems (TAES)

In the SOU, the Edge Technology approach is expressed by TAES—a program design, delivery and analytics system for higher education. The tools enable comprehensive program design and delivery from mapping graduate attributes and standards through building and delivering assessment tasks and courses all the way to the development of an accreditation submission. They make possible a migration from *documenting* the University program development process to a genuine curriculum design paradigm focused on creating high-quality learning experiences. They are different from learning management and other e-learning systems because their functionality is a direct and complete expression of a university's SOU design. The examples described in this chapter come from a suite of three toolkits for higher education learning and teaching (Bain 2012). TAES is comprised of four components.

6.4.2.1 Tools for Program Design: The ProgramSpace

The ProgramSpace includes modules for program design including program conceptualization, standards mapping, program outcomes, assessment task, course and unit design. Figure 6.7 describes a layout employed for standards mapping in the ProgramSpace.

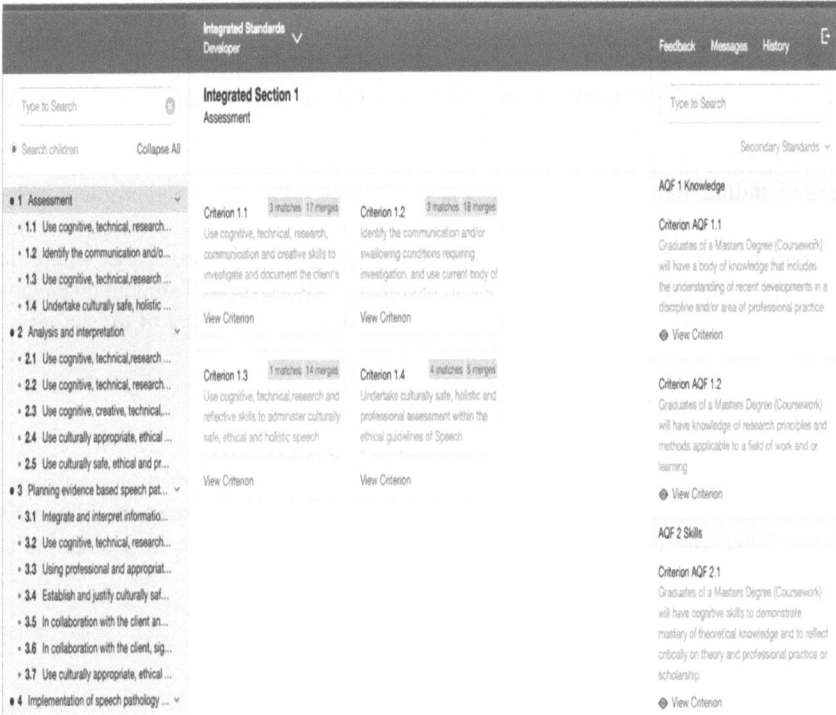

Fig. 6.7 Standards mapping in the Programspace

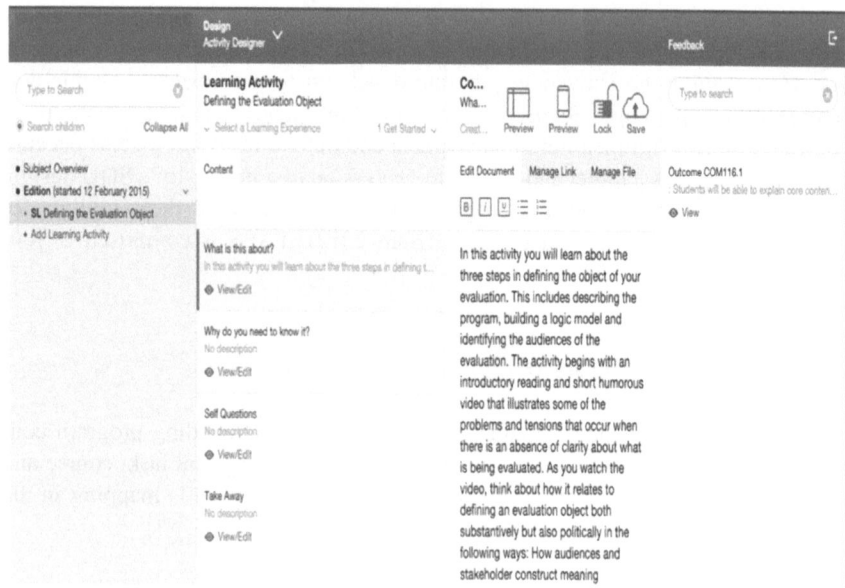

Fig. 6.8 Building an activity-based learning experience in the Coursespace

6.4.2.2 Tools for Course Design and Enactment: The CourseSpace

The CourseSpace includes tools for designing and enacting course learning activities, including tools for building the smarter lectures, cooperative and problem-based learning. Figure 6.8 describes a layout for building an activity-based learning experience. The layout shows the steps in getting started with the design where the designer frames out what the activity is about, what the expected student takeaways are and questions the students could reflect upon at the beginning of the activity.

6.4.2.3 Student Tools: The LearningSpace

The LearningSpace is where students engage using computer, tablet or smartphone technology with the learning experiences developed in the CourseSpace. Figure 6.9 shows what the design described in Fig. 6.8 would look like in a smartphone application for students.

6.4.2.4 Accreditation Tools: The AccreditationSpace

The AccreditationSpace makes possible the online submission of an accreditation document to meet national or state requirements or those of a professional body. Figure 6.10 shows a matrix from the AccreditationSpace software.

Fig. 6.9 Learning experience
on a smartphone

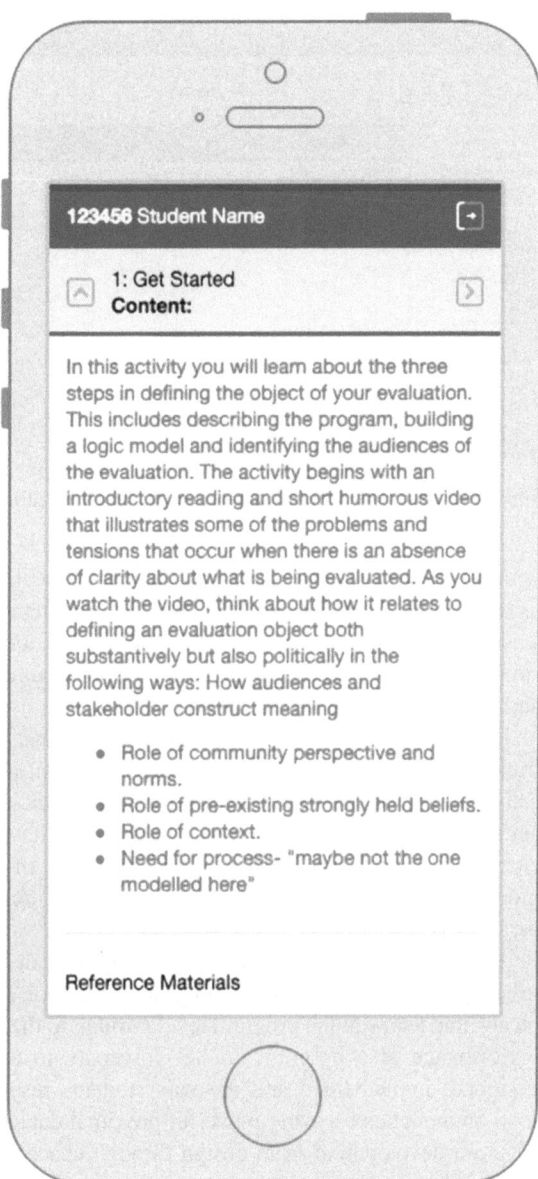

On the left hand side of the layout are the standards set by a national body the three right hand columns show the way evidence is presented to show that those standards are met. The software also makes possible online review of a program submission.

All four "spaces" are connected meaning that a student faculty member or accreditor can look at all aspects of the course and program design as they engage

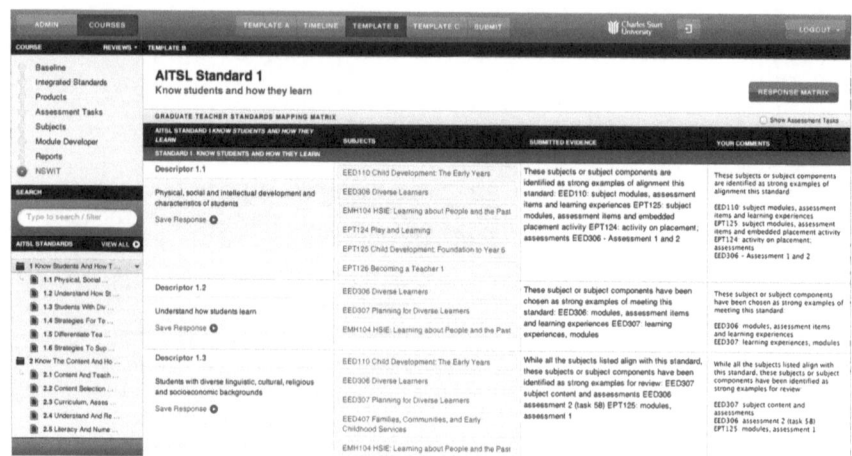

Fig. 6.10 AccreditationSpace standards and evidence matrix

with a learning activity making connections with course outcomes, assessment tasks, and the way the activity fits into the overall program. A faculty member teaching a course can look forward to the student tools and back to the program and course spaces to situate each class session within the context of the way the course and program was designed.

Each toolkit generates emergent feedback that is integrated into an analytics framework for design enactment and engagement at program, course and learning activity level. Figure 6.11 shows the way feedback is recorded in the ProgramSpace and Learning Space. Users answer questions that reflect the key research-based characteristics associated with a particular part of the program or course design process. The figure also shows what feedback might look like from a student's perspective.

The data for any activity, course or program can scale up to reflect data on many learning activities, courses and programs at school, faculty or University level. This means that teams at the program level through to the University level can look at the performance of a program or the University in terms of the way a program is designed, implemented and the way students respond. This approach closes the loop on analytics by using powerful proximal data to generate an understanding of program development from design through enactment to student performance and course accreditation.

In combination, the TAES Spaces create a technology for higher education that is analogous to those that have exerted transformational effects in other fields like business, medicine and engineering. In those fields, technology has been deployed in a manner that is deeply integrated with their core activities or transactions. When a surgeon uses an arthroscope to trim a cartilage, a structural engineer uses computer-assisted design software to simulate the stresses on a bridge, or a sales manager uses customer-relations-management software to predict future inventory

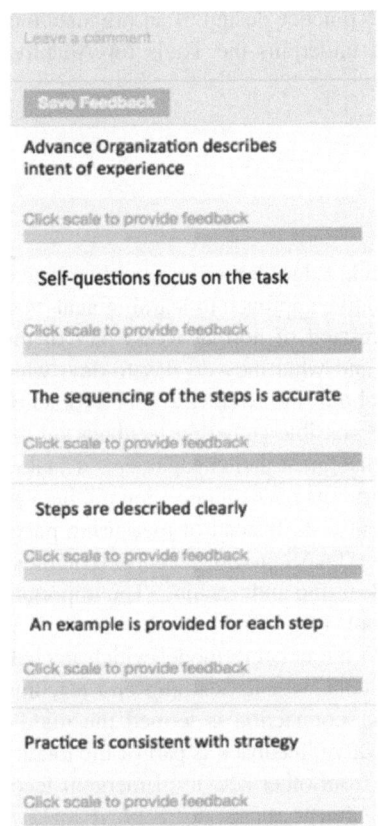
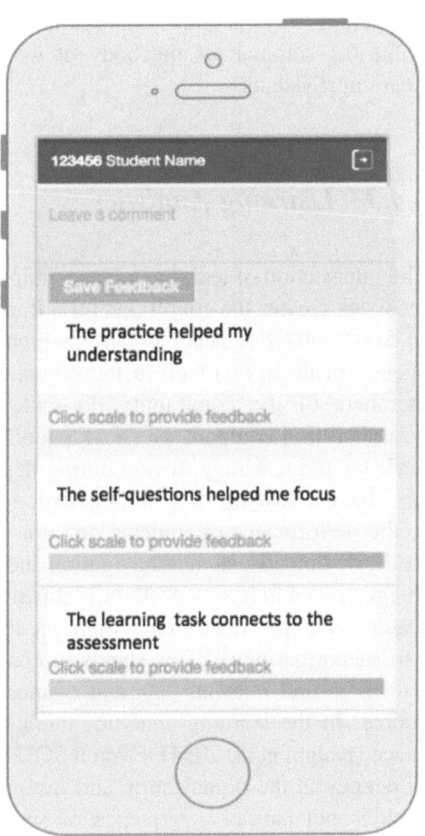

Fig. 6.11 Feedback from faculty and students

needs they do not think about technology. Each one thinks about her or his professional transaction and builds capacity while using the tools (Weston and Bain 2010). They are the venues for normal work.

In the SOU approach TAES spaces are the venues where the organizational design plays out in day-to-day normal work. The spaces represent the practical manifestation of the design. They are where the policies, roles, governance model, and organizational design are reflected in the way work gets done. When technology enables, empowers, and accelerates a profession's core transactions, the distinctions between computers and other devices and professional practice evaporate. No equivalents of these deep and proximal technologically enabled transactions exist within the prevailing educational paradigm of higher education. TAES is designed to generate this equivalency by assisting members of community to do the professional work of learning and teaching.

TAES reflect a body of research dating back to the 1990's that relate to the way educational research and practice can be embedded in learning and teaching software in ways that also express the theory to practice design of an organization. Table 6.1 summarizes the body of work that underpins the Tools for Adaptive Learning Systems.

6.4.3 Learning Analytics

The intersection of learning and teaching research, Edge Technology, and emergent feedback creates the conditions for a transformative approach to learning analytics. In SOU, analytics becomes an emergent expression of normal work, part of the agency of those involved in the organization and what they do day-to-day; what members of the community do with and about the well-researched practice described in this report. The TAES tools make possible gathering feedback on the learning and teaching process during design, enactment and engagement. It makes possible the analysis of a learning experience from its conceptualization all the way to the performance of students on an assessment task. Instead of measuring page-loads, downloads, student navigation and tweets, TAES provides feedback on those things known to have a profound influence on learning and teaching. The important features are embedded in emergent feedback tools (e.g., Fig. 6.10) that are used as part of normal work. This stands in contrast to the existing model which is challenged to find relevant data and connect that data with its learning and teaching source. In the learning analytics literature, this connection is termed the middle space (Knight et al. 2014). From a SOU perspective, feedback is part of the totality of agency in the organization and inseparable from other activity. Emergent feedback is not part of a sequence or space in a linear sense, nor does it make a traditional temporal distinction among data gathering, analysis and reportage.

Issues associated with surveillance, time-delayed ex-post facto feedback, data utility and ethics that are common in critiques of existing analytics approaches (Slade and Prinsaloo 2013) are addressed in the SOU because those involved in the creation and use of feedback are actively engaged with it all of the time. They create it, engage with it, understand it, and use it routinely. No one is surveilled, nor is behaviour subject to ex post facto reportage.

A second major issue in the learning analytics field pertains to what to measure. Most efforts are not informed by the learning and teaching efficacy research described in this report. Data is often gathered on user behaviour including navigation patterns, downloads, tweet analysis, and page loads that have not been shown to influence learning. The SOU, informed by known relationships and efficacy research in learning and teaching, takes up those factors that influence learning in the design of its emergent feedback model. Most important, the learning and teaching potential realised by TAES is an expression of the way theory and research are reconciled in an organizational design that reflects the application of each of the six design principles. A detailed account of the SOU analytics approach is described by Bain and Drengenberg (in press).

Table 6.1 Edge technology research

Question/Area	Study	Method	Result
Gender effects in a 1:1 technology environment	Bain et al. (1999)	Ex-post-facto comparison group design comparing technology immersion program to a control	Discusses the effect of participation in a high computer-to-student access ratio on female student performance, and reports statistically significant skill increases for male and female students in the high access immersion (embedded design) program
Description of curriculum authoring tools	Bain and Huss (2000)	Descriptive article describing antecedent application of Edge technology	Tools employed to build over 2000 h of differentiated curriculum
Initial test of Edge technology principles focused on embedding research-based practice in Information and Communication Technology (ICT) tool design	Bain et al. (2000)	Alternating treatments design, involving use of a hypertext discussion tool to augment research-based teaching	Statistically significant improvement in student achievement when technology employed to augment research-based teaching ($p = 0.003$)
Does use of the curriculum authoring tools change practice	Bain and Parkes (2006a)	Comparison group design- high and lower level technology users to establish whether higher quality teaching practice co-varied with use of a suite of curriculum design tools	Teachers who made higher level use of the technology taught more successfully at statistically significant levels ($p = 0.02$) using the teaching approaches that were part of the Self-organizing School (SOS) design
Do models of knowledge management work in education?	Bain and Parkes (2006b)	Describes issues related to the type and use of data for analytic purposes in education	A foundation of the case for emergent feedback. Introduced concept of proximal and distal data that is a factor in the emergent feedback approach
To what extent was technology used more or less effectively in settings identified as leaders in the use of ICT?	Weston and Brooks (2008)	Comparative case study of four leading US settings (including the Self-Organizing School) identified for the innovative use of technology. Gathered data by survey, interview and school records analysis	More effective technology use at scale in the SOS than all others, deeper faculty understanding and capacity

(continued)

Table 6.1 (continued)

Question/Area	Study	Method	Result
What is the design metaphor for developing emergent feedback tools?	Bain and Swan (2011)	Descriptive article showing the Edge Technology design approach for developing emergent feedback tools	Process used to develop tools employed in the Self-organizing School project
What is Edge technology?	Bain and Weston (2012)	A book describing the Edge approach to technology use in schools	The Edge approach explained in detail with examples
Effectiveness of Edge Technology tools to improve the quality of teaching	Weston and Bain (2014)	Matched-comparison, repeated-measure for intact groups design investigating the mediating effect of a suite of software on the quality of classroom instruction provided to students by teachers. The quality of instruction provided by teachers in the treatment and control groups was documented via observations that were conducted by an independent research team at pre, mid and post intervals of a 225-day study period	No statistically significant differences at pre-test between treatment and control groups. Over three occasions, no statistically significant differences were found for the control group. Statistically significant differences were found for the overall treatment group at mid and post intervals. Moreover, overall differences between the control and treatment groups were statistically significant at mid and post treatment including a post treatment effect size of 1.54. These results suggest that improving instructional quality could be addressed by information and communications technology effectively mediating research and practice
Role and effectiveness of feedback at scale when organizations use Edge Technology	Bain and Weston (2015)	Matched-comparison, repeated-measure for intact groups design of the effect of type, quality and amount of feedback on the quality of instruction by teachers	Preliminary findings show feedback quality and type co-varied with teaching quality at scale

References

Bain, A. (2012). *Smart Tools (Versions1.0 and 2.0) Computer Software*. Bathurst, NSW: Charles Sturt University.

Bain, A., & Drengenberg, N. (in press). *Transforming the measurement of learning and teaching in higher education*. New York: Routledge.

Bain, A., Hess, P., Berelowitz, C., & Jones, G. (1999). Gender differences and computer competency: The effects of a high access computer program on the computer competency of young women. *International Journal of Educational Technology, 1*(1). Retrieved from http://www.ascilite.org.au/ajet/ijet/index.html.

Bain, A., & Huss, P. (2000). The curriculum authoring tools: Technology enabling school reform. *Learning and Leading with Technology, 28*(4), 14–17.

Bain, A., Huss, P., & Kwong, H. (2000). The evaluation of a hypertext discussion tool for teaching English literature to secondary school students. *Journal of Educational Computing Research, 23*(2), 203–216.

Bain, A., & Parkes, R. J. (2006a). Curriculum authoring tools and inclusive classroom teaching practice: A longitudinal study. *British Journal of Educational Technology, 37*(2), 177–190. doi: 10.1111/j.1467-8535.2005.00527.x.

Bain, A., & Parkes, R. J. (2006b). Can education realize the potential of knowledge management? *The Canadian Journal of Learning and Technology, 32*(2).

Bain, A., & Swan, G. (2011). Technology enhanced feedback tools as a knowledge management mechanism for supporting professional growth and school reform. *Educational Technology Research and Development, 59*(5), 673–685. doi:10.1007/s11423-011-9201-x.

Bain, A., & Weston, M. E. (2012). *The learning edge: What technology can do to educate all children*. New York: Teachers College Press.

Bain, A., & Weston, M. (2015). *The scalable effects of feedback on the quality of teaching: A technology mediated approach*. Manuscript in preparation.

Barabási, A. (2002). *Linked: The new science of networks*. New York: Perseus.

Baume, D., & Yorke, M. (2010). The reliability of assessment by portfolio on a course to develop and accredit teachers in higher education. *Studies in Higher Education, 27*(1), 7–25. doi:10.1080/03075070120099340.

Knight, S., Buckingham Shum, S., & Littleton, K. (2014). Epistemology, assessment, pedagogy: Where learning meets analytics in the middle space. *Journal of Learning Analytics, 1*(2) Retrieved from http://epress.lib.uts.edu.au/journals/index.php/JLA/article/view/3538.

Slade, S., & Prinsaloo, P. (2013). Learning analytics: Ethical issues and dilemmas. *American Behavioral Scientist*. doi:10.1177/0002764213479366.

Weston, M. E., & Bain, A. (2010). The end of techno-critique: The naked truth about 1:1 laptop initiatives and educational change. *The Journal of Technology, Learning, and Assessment, 9*, 6.

Weston, M. E., & Bain, A. (2014). Bridging the research-to-practice gap in education: A software-mediated approach for improving classroom instruction. *British Journal of Educational Technology*. doi:10.1111/bjet.12157.

Weston, M. E., & Brooks, D. (2008). Critical constructs as indicators of a shifting paradigm in education: A case study of four technology-rich schools. *Journal of Ethnographic and Qualitative Research in Education, 2*(4), 281–291.

Chapter 7
Implementation

Abstract This chapter describes the phases and stages of implementation of the self-organizing University including the major needs and challenges associated with each phase. This includes an expected time frame and a description of the work that needs to be completed in each stage focusing on the role of leadership in the process.

Keywords Baseline · Initiation · Design · Scale-up · Pilot · Consolidation · Schema · Transformational change · Adoption · Leadership · Inclusive

The SOU assumes a five to seven year implementation cycle that includes the following four phases:

7.1 Baseline Assessment and Design Phase

The SOU begins with a community baseline assessment and design year. The baseline and design year accomplishes two things. It makes the case for the change and readies the community for the process. It creates the policy and organizational foundation to make the change happen. This includes identifying policy gaps (e.g., lack of a policy on learning and teaching), policy alignment (e.g., revising existing polices to align with the key elements of the change), designing the governance infrastructure (e.g., creating a self-similar and distributed team approach to decision-making) and developing the communication and participatory design approach to ensure that the baseline and design is undertaken by the community not done to it. In this phase, current needs are instantiated and policies, processes, tools and plans are developed for a new model. During the baseline and design year approval is secured from the university's governance to proceed to a pilot phase. The necessary developments and changes are made to embed the new approach in the university's normal operation for program, design approval and implementation.

© The Author(s) 2016
A. Bain and L. Zundans-Fraser, *Rising to the Challenge of Transforming Higher Education*, SpringerBriefs in Education,
DOI 10.1007/978-981-10-0261-8_7

Key challenges in this phase are first, ensuring broad-based exposure to and engagement with the needs and drivers for the change. The understanding of these needs and drivers are a first step in schema building. Engagement is a particular challenge in the early stages of a major change where the findings of a needs assessment and the work of policy development may seem conceptual and abstract, unrelated to the actual day-to-day work of the community. Second, and related to the first challenge, is the need to create a participatory and collaborative design process that ensures the broader community is comprehensively represented in the design of those new policies and processes. To do so, requires applying the self-similarity and dispersed control principles of the theory to build a matrix of stakeholder groups and working teams that will both design the changes and assume the ownership required to disseminate the design work to their respective constituencies.

The critical risk in this phase is insufficient or superficial engagement with the change that weakens support moving forward. As such, it is essential for leadership to *make* and *keep making* the case for the needs and drivers associated with the change and to demonstrate active engagement with the design process. This must be the case for leaders at all levels of the existing organization. Every community requires direction in a big change. The signals for a direction change emerge from a commitment to the effort in ways that clearly and continually foreshadow the changes a new model will bring. All leaders need to communicate the results of the baseline and design work knowingly and authoritatively in order to amplify commitment to the process.

7.2 Initiation Phase

The initiation phase of the SOU is a 12–15 month pilot that establishes and tests out the complete design with a small number of early adopting faculty teams and programs. The size of the pilot is determined by the organization although the focus at this stage should be on implementing a complete design with integrity. The pilot identifies needs, issues and successes that shape the approach required to bring the model to scale across the university. Most important in this phase is ensuring that the key elements of the design (e.g., governance, role, technology and professional support) all emerge from the design phase in a form that are complete and can be reliably tested in a pilot process. Feedback and management are critical in this phase. The governance design and its teams are the venues for leadership and faculty do the active work of guiding and managing the process proactively, supporting participants, and importantly sharing feedback in ways that both address issues in a timely manner and inform the refinement of the design.

Strong and steady communication needs to occur in this phase within the project as well as to the broader community. Leaders require the expertise to test feedback against the intents and expectations of the baseline assessment and design, showing that the project overall is listening to the community while also pushing forward assertively on action that relates to the agreed upon key purposes and rationale. The latter is a significant challenge given that in many higher education organizations leaders may or may not have extensive background experience and capacity with the new approach and its learning and teaching underpinnings. Those leaders may be learning about the full scope of the changes at the same time as the broader community. It is important to recognize that this lag can result in diminished engagement of leadership especially at the middle management level and represents a serious risk to the progress of the change. It can be averted by strong communication and an intensive commitment to providing leaders with the solutions and professional growth opportunities in advance of each phase.

A critical related risk in the pilot phase is the *"building the jumbo jet in flight problem"* where two systems are in effect operating in a single organization. This can engender problems in two extremes. First, the pilot effort can be marginalized because of the small scale of the effort; it is seen as a "side line" initiative that does not apply to the whole community. This is an extension of the low engagement problem in the design phase and represents an ongoing problem if the project and its implications race ahead of the community's knowledge, understanding and comfort with the change. The second risk is a corollary of the marginalization problem where over or special attention is focused on the incubation of the pilot at the expense of the majority model. This can result in a perceived devaluing of the traditional core work of the organization because of what is seen as a hyper focus on the initiation phase.

"Busyness" is also a related risk especially for managers who are too busy meeting their existing responsibilities to pay attention to the new approach. Addressing these issues requires careful thought and planning in order to allocate the time required for stakeholders to assume responsibility for their multiple roles associated with an organizational design in transition. In this phase it is also important to find synergies between existing and new structures and look for ways to dovetail "like" initiatives. For example, new requirements in curriculum, teaching, assessment and technology can be folded into the existing portfolio of professional development approaches the university offers to faculty through its existing model of support for learning and teaching (e.g., its center for learning and teaching). This can occur in advance of faculty member participation in the new model.

Leadership of such an important and broad-based change must be inclusive. There is a tendency at the higher levels of universities to make a distinction between strategic and operational leadership. This distinction does not hold for transformational change. Many of the challenges and needs associated with an all-of-organization change fall in the space between strategy and operations as new strategy interacts with new design and normal work. As such, making hierarchical leadership distinctions between roles for strategy and operations are generally unproductive in a large-scale change. Transformational change requires all leaders at all levels to be actively involved across the strategy/operation divide. Any retreat from commitment in this phase can place a new approach at serious risk.

7.3   Scale-up Phase

The SOU approach is based on a migration methodology (Bain 2007) whereby the university creates a new model within its existing operation and then migrates the organization to that model after securing successful initial design, piloting and preliminary scale-up. Feedback in all phases is critical to determine the efficacy of the initial work and to inform subsequent design and implementation decisions. The scale-up phase is also flexibly determined based on the result of the pilot and occurs in waves bringing programs and/or faculties on board at a pace consistent with the organization's appetite for change, leadership capability, cost factors, and the findings of the pilot implementation.

The critical risk in the scale-up phase is the misalignment of the rate of adoption and the support required to successfully scale the approach. For example, pressure to get the model in place quickly in order to achieve institutional expectations or outcomes (e.g., better performance on competitive external measures of teaching quality) can result in scaling up at a rate that exceeds the organization's capacity to provide professional development and in situ support for faculty members. Those faculty members who have successfully navigated the change provide the best induction to a new model in its scale-up phase. However, scaling up too fast means that those mentors can be overwhelmed by the amount of support required by their newly involved peers as they encounter the just-in-time problems associated with doing learning and teaching differently. Establishing ratios for mentoring and support and using the feedback and experience of early adopters to build *rate of adoption* data can bring a stronger empirical foundation to the effort and inform the critical scale up phase of the SOU approach.

Feedback continues to drive the process in this phase. Software, governance structures and process are all subjected to continued refinement as the governance model and its self-similar and distributed structure of groups and teams drive the constant feedback necessary for self-organizing change and adaptation. For example, the scale up phase will provide the important ongoing feedback necessary for refining the TAES software platform that drives the SOU and may result in a new version. That version will reflect and drive improvement in the overall program design and delivery process. It is in this phase that some of the time dependent features of the overall design begin to exert an effect. For example, in the scale-up phase some faculty members will have accumulated sufficient experience with their new roles to seek promotion using a new promotion framework that reflects the values and skills associated with the new model. Leaders who in the initiation phase of the process have no more experience than the people they supervise will have the time and opportunity to build demonstrable new skills and leadership capacity consistent with expertise in the model. Both the career trajectory and instantiation of the leadership roles required to consolidate the model can enhance schema building and the extension of the model at scale. It is also normal to expect an increase in the turnover of faculty and administrative staff in this phase as the new model begins to take hold and members of the community evaluate whether they want to come on board with the new approach.

All involved in a transformational change process need to recognize and expect the angst and pushback that is normal with a new all-of-organization approach to learning and teaching, especially when former practice has been largely private and autonomously defined. This angst is often amplified in the scale-up phase. Executive leadership must continue to echo and reiterate the need and rationale for the new approach as defined in the baseline assessment and design phase and assume responsibility for all aspects of the process. Most important, is to acknowledge and respect all feedback and work with it. In doing so leaders must have enough engagement and knowledge to analyse, filter, and respond to that feedback in ways that enable it to be both responsive and managed effectively. Any expectation that an all-of-organization change can occur without disruption or that simply building a change management infrastructure will avert disruption is naïve. All involved should be aware of and expect periods of high challenge where consistent and frequent communication is key and where the responses provided to questions will not always be viewed by all constituencies and stakeholders as satisfactory.

7.4 Consolidation

In the consolidation phase the process of migration is completed with the new organizational design completely replacing the old. This phase signals the end of the dual approach in the organization and the full implementation of all aspects of the new design. The organization is no longer in a major change process although the theory in practice invites constant change as a hallmark of a dynamic adaptive organization. Nonetheless, the major milestones in accomplishing the change to a SOU approach is completed in the consolidation phase. This includes phasing out older committee and support structures, embedding all ancillary approaches to professional support within the model, and fully activating the promotion and leadership models based upon advancing expertise. At this time there is also sufficient capacity within the organization to readily provide the mentoring support and professional development to meet the normal cyclical turnover in faculty and administrative staff. In this phase the university can expect to be self-organizing where the self-similarity and dispersed control principles expressed in the organizational design is fully informing the new approach to learning analytics; there is a genuine organizational schema for learning and teaching and leadership can lead with the knowledge and expertise required to respond to the feedback that emerges from the normal work of program and course design at scale. The community in this stage possesses the agency associated with the expertise that makes the new model a genuine expression of the normal work of a learning community that possesses its unique understanding of learning and teaching at scale.

The greatest challenge in this phase is regression where the organization despite the hard work and success of its change engages in equilibrium seeking behavior that draws it back to old practice. The best response to this risk is continued focus on the feedback generated at all levels of the organization that informs dynamic change and the evolution of the schema for learning and teaching. To do so involves continuing to evolve the tools, methods and governance process that supports the core activity, the design, enactment and engagement with the university's model of learning and teaching.

Reference

Bain, A. (2007). *The self-organizing school: Next-generation comprehensive school reforms.* Lanham, MD: Rowman & Littlefield.

Chapter 8
The Design in Practice

Abstract In this chapter we describe the developmental trajectory of the self-organizing university in practice, the way the model was developed and its elements tested in a range of settings.

Keywords Self-organizing school · Complexity theory · Hong Kong schools · Pattern language · Schema development · Reform

The SOU design approach described here was first tested in the self-organizing school project (Bain 2007), a longitudinal field-based experience with principles and practices derived from an 11-year secondary-school reform project. The main intent of the project was to test the design approach by creating a reform process and model that embedded innovative practices and processes that influenced student learning within and across the school system (Bain 2007, p. 5). From a student perspective, the project was designed to improve current levels of performance through a program that combined best practice in teaching, curriculum, and resources. Criticism of previous reform attempts had questioned whether it was possible to embed and sustain effective teaching and learning practices in schools. The self-organizing school responded to this by sustaining a design that offered theory, practice and evidence demonstrating the comprehensive nature of the design and implementation approach. The theory and methodology used for this project were deeply embedded in complexity theory, complex adaptive systems and self-organization.

Within the self-organizing school (Bain 2007) the mechanisms for sharing feedback were embedded within the design and were central to its implementation and ongoing success. Individual performance was noted over the whole time this system operated, as the natural cycle of communication set up within the system ensured that feedback was provided continually rather than just at critical times when decisions needed to be made. This feedback was emergent and formative rather than summative. The school's design and schema created the opportunity for all members of the team (teachers and students) to set goals and to regularly share their growth towards these goals. In this manner the agents involved within that

© The Author(s) 2016

A. Bain and L. Zundans-Fraser, *Rising to the Challenge of Transforming Higher Education*, SpringerBriefs in Education,
DOI 10.1007/978-981-10-0261-8_8

system embedded the feedback process in a way that supported and developed the broader purposes not only of the team but of the school as well.

8.1   Further Traction

Beyond the investigation of the application of principles of self-organization in the self-organizing school project, additional traction for this design has been found through the Hong Kong school project (Bain et al. 2011). That project used a design built around the six principles of self-organization and targeted capacity-building in a Hong Kong co-educational secondary school, begun in 2009 and ongoing. There were four strategic goals for the project: to integrate the principles of the self-organizing school with the school's vision and goals; to apply the principles to building school capacity in mastery teaching, cooperative learning, differentiated instruction, and the use of technology; to develop and use a lesson plan database to help build a school repository; and to build a system within the school for recognizing and rewarding teachers, including promotional opportunities and promotion criteria that aligned with the school goals. At the time of publication (2011) the project was travelling well towards these goals, with explicit acknowledgement of the importance of leadership, clear vision and goals and the need to create conditions where this type of innovation can occur.

8.2   Application of Self Organization in Higher Education

The design principles have also been subject to additional application in the higher education context in a decade long program of research that is ongoing and focused on pre-service and graduate teacher preparation (e.g., Bain et al. 2009; Zundans-Fraser and Auhl 2015; Zundans-Fraser and Lancaster 2012; Zundans-Fraser 2014; Zundans-Fraser and Bain 2015). This work has examined the principles as a program design and evaluation heuristic, in student pattern language and schema development, content knowledge acquisition, and research to practice skill development. The findings derived from studies in both the compulsory and higher education sector are summarized in the following section describing efficacy research.

References

Bain, A. (2007). *The self-organizing school: Next-generation comprehensive school reforms.* Lanham, MD: Rowman & Littlefield.

Bain, A., Lancaster, J., & Zundans, L. (2009a). Pattern language development in a pre-service learning community. *International Journal of Teaching and Learning in Higher Education, 20* (3), 336–349.

Bain, A., Lancaster, J., Zundans, L., & Parkes, R. J. (2009b). Embedding evidence-based practice in pre-service teacher preparation. *Teacher Education and Special Education, 32*(3), 215–225. doi:10.1177/0888406409339999.

Bain, A., Walker, A., & Chan, A. (2011). Self-organization and capacity building: Sustaining the change. *Journal of Educational Administration, 46*(6), 701–719. doi:10.1108/09578231111174839.

Zundans-Fraser, L. (2014). *Self-organisation in course design: A collaborative, theory-based approach to course development in inclusive education.* (Unpublished doctoral dissertation). Charles Sturt University, Bathurst.

Zundans-Fraser, L., & Auhl, G. (2015). *A theory-driven approach to course design in inclusive education.* Manuscript submitted for publication.

Zundans-Fraser, L., & Bain, A. (2015). The role of collaboration in a comprehensive programme design process in inclusive education. *International Journal of Inclusive Education.* Advance online publication. doi:10.1080/13603116.2015.1075610.

Zundans-Fraser, L., & Lancaster, J. (2012). Enhancing the inclusive self-efficacy of pre-service teachers through embedded course design. *Education Research International, 2012.* doi:10.1155/2012/581352.

Chapter 9
Efficacy Research Underpinning the SOU

Abstract This chapter describes the prior applied work and research that underpins the self-organizing university as a whole including early research in the compulsory sector and the transition of the model to higher education. Key studies are identified including a description of their contribution to the overall research base for the model.

Keywords Achievement · Practice fidelity · Higher education · Student learning · Self-efficacy · Embedded design · Schema development · Pattern language

As with any theory, the central question about its utility is whether it works in relation to the problem or challenge it seeks to explain or resolve. The research on the application of the self-organizing design principles to education has occurred over a 22-year period and continues through the SOU.

9.1 The Self-organizing Schools Project

The cornerstone of the body of work underpinning SOU is the 12-year comprehensive research-practitioner study the "self-organizing school (SOS) project" (described earlier) that established and tested the principles described in Table 9.1 in the re-design of a secondary education setting (Bain 2007). A number of authors have indicated that higher education needs to look to the longer history of reform in K-12 education for insights and a foundation for innovation (Lam and Tsui 2014; Wang 2014). The SOU has a foundation in K-12 research.

The Central Question for the self-organizing school project was: Could a complex educational organization employ the design principles to reform its design and learning and teaching processes in ways that were more responsive to students' learning needs and resulted in better teaching, learning and achievement at scale? The central question served as a guide for the development of more specific sub-questions related to the integrity of implementation of the design, effects on student learning, levels of collaborative practice, and the use of technology.

© The Author(s) 2016
A. Bain and L. Zundans-Fraser, *Rising to the Challenge of Transforming Higher Education*, SpringerBriefs in Education,
DOI 10.1007/978-981-10-0261-8_9

Table 9.1 SOU efficacy research from the self-organizing school project

Question/Area	Study	Method	Result
Did improvement in student achievement co-vary with the introduction of the design?	Bain and Ross (1999)	A longitudinal cohort design to compare SAT-1 performance for 160 students prior to the change process with those who had participated in the theory-based change. Entry profiles used as a covariate comparison indicated no significant difference at entry	92 point improvement in SAT-1 performance differential ($p = 0.0003$) in favour of the students who participated in the change. Also, established gains for students with a learning disability
Did the design create a collaborative process and culture?	Bain and Hess (2002)	Comparison cohort design study based upon self-organizing school (SOS) pre and during the theory—based change process. Dependent measure—survey data gathered in situ and compared with data derived from 42 other like schools (aggregated on the same survey)	Positive reports of collaborative culture at higher levels than pre-change in the target school and than in the data from comparison schools
Did theory and the design principles result in an altered model of school design and was it implemented with integrity?	Bain (2007); Weston and Brooks (2008)	Five year longitudinal case study involving dependent measures of over 1600, 55 min classroom observations, triangulated with over 1300 teacher, management and team surveys and 12,000 student evaluations	Successful implementation of the design based upon the theoretical principles as evidenced by change in pedagogical practice and high levels of implementation integrity of that practice, sustained over five years and triangulated with teacher, team, student, and management surveys perspective. High levels of implementation integrity over five years indicating sustainability of the design

(continued)

Table 9.1 (continued)

Question/Area	Study	Method	Result
What was the practice fidelity of the SOS project?	Bain (2010)	Regression analysis of four years of formal observations of classroom teaching practice (criterion variable) with the perspectives of administrators, teachers, and teaching peers about the reform's implementation to ascertain the practice fidelity of the SOS	The results showed sustained levels of practice fidelity and statistically significant differences in the ratings of administrators, teachers and peers, although those differences reduced overall as the reform progressed
Application of the theoretical principles to planned change in Hong Kong	Bain et al. (2011)	Descriptive case study employed to illustrate the SOS principles and discuss their role in site-based capacity building	Describes the way the principles were taken up in a high achieving secondary school setting
Use of the theoretical principles to determine the practice fidelity of a program for gifted students in a Canadian elementary school	Boyd (2012)	Mixed methods case study approach	Showed a moderate to high-level of practice fidelity related to the quality of implementation of the program

Table 9.1 summarizes the program of research and specific questions asked in the self-organizing schools project and the results of empirical study.

When viewed together, the studies both qualitative and quantitative indicate that the SOS design principles did influence the school design, were able to exert an influence on the quality of teaching and learning in the self-organizing school in sustainable ways. This included demonstrable improvement in student achievement, the quality of teaching, levels of collaboration and collaborative culture, and the use of technology described subsequently. Further, the application of the approach in primary and secondary school settings suggest its utility both as a design metaphor and an evaluative heuristic for determining practice fidelity.

Table 9.2 SOU efficacy research in higher education

Question/Area	Study	Method	Result
Schema development as reflected in pattern language acquisition	Bain et al. (2009a)	Uninterrupted time series design with 54 volunteer pre-service teacher educators. Dependent measure—use of pattern language	Improvement in frequency and quality of student pattern language use co-varied with the application of the embedded design principle in the course learning experience
Student achievement in a course designed employing the embedded design principle	Bain et al. (2009b)	Repeated Measures counterbalanced design with 90 volunteer pre-service educators employing an achievement quiz	Mastery level knowledge co-varied with the application of the embedded design principle
Effect of the design on self-efficacy	Lancaster and Bain (2010)	Comparison group design involving 36 pre-service educators using self-efficacy measure (SEIPD)	Statistically significant improvement in student achievement for both conditions-one involving applied experience, the other embedded design
Research to practice	Grima-Farrell (2012)	Comparative case study mixed methods design employing surveys, focus groups and permanent product records	Linked course design to research to practice capability of graduate educators
The use of embedded design to enhance student self-efficacy	Zundans-Fraser and Lancaster (2012)	Comparison group design involving 41 pre-service educators using self-efficacy measure (SEIPD)	A theoretically designed course significantly improved self-efficacy between pre and post occasions
Applicability of the design principles to higher education. Can the approach be employed purposefully to guide a program design process?	Zundans-Fraser (2014)	Eight-year longitudinal case study employing design-based research approach. Measures include semi-structured interviews, researcher's log and observational notes, document analysis	Validated that the key issues identified in the literature were present in the study setting. Demonstrated that the principles could be employed as a design metaphor in HE program and course design. High levels of satisfaction expressed by the design team, stakeholders and students

(continued)

Table 9.2 (continued)

Question/Area	Study	Method	Result
Schema development	Auhl (in progress)	Longitudinal comparison group design examining the ways in which schema development co-varied with the type of preparation program. Dependent measure schema development	Schema for professional practice was most developed in a program designed using the SOS principles
Effect of program design on performance in workplace learning setting	Lancaster (in progress)	Comparison group design using direct observation of teacher led instruction using an observational scale	Students who experienced the embedded design approach taught more effectively in situ during a practicum experience

9.2 From School to University

A key question often asked about the SOU pertains to its trajectory from the initial work in schools to its application to learning and teaching in higher education settings. This body of research began in 2003 and is continuing. It involved applying the principles to the design of graduate level programs in inclusive education as a way to address the concerns about higher education program design identified earlier in this report. The cornerstone of that effort is a recently completed study by Zundans-Fraser (2014), describing the application of the six theoretical principles to the design of a higher education program. The study described a design-based research approach to the challenges faced by universities in program design and articulated a process to respond to those challenges using the theoretical principles described here. This initial work has expanded to include studies of the effects of the design approach on professional schema development (Auhl, in progress) student learning, self-efficacy and pattern language (Bain et al. 2009a, b; Lancaster and Bain 2010; Zundans-Fraser and Lancaster 2012), in situ teaching to practice of graduate teachers (Lancaster, in progress), the theory to practice work of graduate teachers (Grima-Farrell 2012), and organizational design in higher education (Zundans-Fraser and Bain 2015a, b). Table 9.2 summarizes the program of research in higher education and the results of empirical study.

The studies described in Table 9.2 lend empirical support to the SOU as an effective higher education curriculum design methodology that responds to the key areas of need identified in the literature. The results also echo the findings of the original self-organizing school project showing that the design approach, improves achievement, teaching outcomes, student disposition, schema development and the translation of research to practice. Further, the work of Zundans-Fraser (2014) provides a methodological framework for the use of the theory in a higher education setting.

References

Bain, A. (2007). *The self-organizing school: Next-generation comprehensive school reforms.* Lanham, MD: Rowman & Littlefield.

Bain, A. (2010). A longitudinal study of the practice fidelity of a site-based school reform. *Australian Educational Researcher, 37*(1), 107–124.

Bain, A., & Hess, P. (2002). *School reform and faulty culture: A longitudinal case study.* (ERIC Document Reproduction Service No. ED472655.

Bain, A., Lancaster, J., & Zundans, L. (2009a). Pattern language development in a pre-service learning community. *International Journal of Teaching and Learning in Higher Education, 20* (3), 336–349.

Bain, A., Lancaster, J., Zundans, L., & Parkes, R. J. (2009b). Embedding evidence-based practice in pre-service teacher preparation. *Teacher Education and Special Education, 32*(3), 215–225. doi:10.1177/0888406409339999.

Bain, A., & Ross, K. (1999). School re-engineering and SAT-1 performance: A case study. *International Journal of Educational Reform, 9*(2), 148–154.

Bain, A., Walker, A., & Chan, A. (2011). Self-organization and capacity building: Sustaining the change. *Journal of Educational Administration, 46*(6), 701–719. doi:10.1108/09578231111174839.

Boyd, C. (2012). *Practice fidelity and sustainability of school reform: A study of the school wide enrichment model in an independent elementary school.* (Unpublished doctoral dissertation). Charles Sturt University, Bathurst.

Grima-Farrell, C. (2012). *Identifying factors that bridge the research-to-practice gap in inclusive education: An analysis of six case studies.* (Unpublished doctoral dissertation). Charles Sturt University, Bathurst.

Lam, B. H., & Tsui, K. T. (2014). *Curriculum mapping as deliberation—examining the alignment of subject learning outcomes and course curricula Studies in Higher Education.* doi:10.1080/03075079.2014.968539.

Lancaster, J., & Bain, A. (2010). The design of pre-service inclusive education courses and their effects on self-efficacy: A comparative study. *Asia-Pacific Journal of Teacher Education, 38* (2), 117–128. doi:10.1080/13598661003678950.

Wang, C.L. (2014). Mapping or tracing? Rethinking curriculum mapping in higher education. *Studies in Higher Education.* doi:10.1080/03075079.2014.899343.

Weston, M. E., & Brooks, D. (2008). Critical constructs as indicators of a shifting paradigm in education: A case study of four technology-rich schools. *Journal of Ethnographic and Qualitative Research in Education, 2*(4), 281–291.

Zundans-Fraser, L. (2014). *Self-organisation in course design: A collaborative, theory-based approach to course development in inclusive education.* (Unpublished doctoral dissertation). Charles Sturt University, Bathurst.

Zundans-Fraser, L., & Bain, A. (2015a). The role of collaboration in a comprehensive programme design process in inclusive education. *International Journal of Inclusive Education.* Advance online publication. doi:10.1080/13603116.2015.1075610.

Zundans-Fraser, L., & Bain, A. (2015b). *How do institutional practices for program design and review address areas of need in higher education.* Manuscript submitted for publication.

Zundans-Fraser, L., & Lancaster, J. (2012). Enhancing the inclusive self-efficacy of pre-service teachers through embedded course design. *Education Research International, 2012.* doi:10. 1155/2012/581352.

Chapter 10
Impacts and Benefits

Abstract This chapter unpacks the benefits that can be expected from applying the self-organizing university including the way better learning and teaching can get to scale in a university, Benefits include higher quality curriculum, sophisticated analytics data, more equitable performance management, and improved student outcomes. The chapter focuses on distinctiveness and the way the self-organizing university model can provide a university with a distinctive empirically supported approach to learning and teaching.

Keywords Attribution · Rankings · Professional development · Distal · Career growth · Professional control · Productivity · Comparable and visible practice · Distinctive

Longitudinal empirical research shows that at the organizational level, universities face immense difficulty differentiating themselves from each other on the basis of their approaches to learning, teaching and student success. (e.g., ACT 2009; Jankowski et al. 2012; Pascarelli and Terenzini 2005). All universities assert an influence on the quality of teaching and student outcomes. At this time, universities experience great difficulty showing how such influence occurs at scale across the organization. Universities claim such effects aspirationally in order to seek competitive advantage, although this occurs without empirical support beyond distal correlational measures of faculty, student and institutional performance (Dvorak and Busteed 2015). A university that applies the SOU approach can expect to alter this circumstance in the following ways:

1. The end of the Black-Box: A university that applies the SOU model builds a deep understanding of teaching and learning in all phases and at the scale of the whole community. It has a schema; faculty and students know and understand what learning and teaching means. Their understanding and commitment is continually informed by feedback linked to an evolving understanding of the learning and teaching process. Instead of waiting and hoping for the outcomes of student experience and satisfaction surveys and national rankings, the SOU can use its proximal data to inform its process all of the time. The key impact

© The Author(s) 2016   69
A. Bain and L. Zundans-Fraser, *Rising to the Challenge of Transforming Higher Education*, SpringerBriefs in Education,
DOI 10.1007/978-981-10-0261-8_10

associated with the "*end of the black box*" is the development of an attributable relationship between what teachers do and student learning that is transparent, understandable, and empirically derived. A university that possesses an attributable understanding of learning and teaching can distinguish itself from its competition in powerful and verifiable ways.

2. From Ancillary to Embedded Support: Nearly all universities have centers responsible for the promotion of quality learning and teaching. Their activity includes: conducting programs for *teaching awards, ongoing professional development* some of which may lead to formal qualifications, *grants for learning and teaching innovation*, and serving as a repository for cutting edge *learning and teaching resources*. It is clear given the research noted in the introduction to this section, that while this type of ancillary approach engages in good and important work, it is not of the scope or scale required to exert an all-of-organization impact on the quality of learning and teaching in universities. The SOU approach offers something different in this regard by embedding the work of supporting learning and teaching as and integral part of the university's model. In this embedded approach supporting learning and teaching becomes part of the organization's design. For example, teaching awards in a SOU approach recognizes those individuals, who demonstrate excellence in the university's model of learning and teaching, who express originality and innovation, and who move the model forward pushing it to new limits and places. Where formerly, those awards were based on distal student satisfaction measures, and often given for initiatives of unsubstantiated efficacy, in SOU, awards are an expression of the emergent feedback about learning and teaching produced constantly by TAES tools through the process of learning design, enactment and engagement. The awards possess agency. They represent an empirically verifiable expression of the university's model and direction; they become part of the overall schema. As well as recognizing excellence, they serve as a catalyst for innovation and development—they drive the evolution of the schema. Similarly, grants and projects seek to address emergent needs associated with the evolution of the university's distinctive approach to learning and teaching while the repository of resources are also aligned with the design and schema. Professional development or capacity building in the SOU approach focuses on a clear understanding of current and projected needs as feedback informs the next evolution of all aspects of the design. Support may be provided by a centre or within faculties although the key distinguishing feature of any capacity building effort is that it is part of the model and organizational design, as opposed to being exemplary of a generic understanding of learning and teaching, juxtaposed or ancillary to normal learning and teaching work.

3. Legitimacy in Career Growth and Learning and Teaching: Comparable and visible quality practice (Bowker and Starr 2000) that can be attributed to learning is an expected outcome of SOU. It also a prerequisite condition for establishing any equitable system of recognition or merit in the learning and teaching domain. The inability to make such a connection at scale, has made the

merit-based recognition of learning and teaching in both the compulsory and higher education sectors of education highly controversial and of questionable efficacy (e.g., Springer et al. 2010; Harrison and Mather 2016). The SOU approach focuses on the kind of rigour and professional control (Bowker and Starr 2000) necessary to provide legitimacy to the learning and teaching career trajectory because it establishes an attributable empirical connection between what teachers do and student learning. By doing so, universities can create the conditions to build greater rigor and esteem in the learning and teaching dimensions of their core activity.

4. New Data: The SOU through the TAES produces learning and teaching data derived from professionally controlled practice that is proximal to those things that make a genuine learning and teaching difference. The capacity to gather genuine learning and teaching data can transform a university's approach to learning and analytics and its understanding of itself.

5. Understanding and Managing Learning and Teaching Productivity: One of the reasons why defining educational productivity remains such a challenge for higher education is the difficulty in defining what the term actually means. Efforts are plagued by a basic inability to measure key inputs and outputs related to what constitutes quality (Massey et al. 2012). Time often becomes the proxy for costs associated with designing programs and courses. Faculty members are allocated time to complete a course or syllabus or build a MOOC based on estimates of how long such an activity should take without giving attention to what constitutes an effective design. The widely disparate estimates of the time required to develop courseware is an indicator of just how problematic this time as proxy turns out to be (e.g., Hollands and Tithali 2014). The time as proxy issue is a by-product of the black box problem described above. Without a model of what constitutes effective practice and a way to attribute that practice to quality student outcomes, there is no way to make meaning of time spent beyond counting successful graduands and credit hours completed.

The SOU offers something different. The application of theory to design and the instantiation of design in practice means that the learning and teaching process in design, enactment, and engagement is understood in terms of comparable and visible practice- normal work. Further, constant emergent feedback shows whether learning and teaching in each phase is effective. So, with the SOU we know what we are doing and whether it works. This means that from an educational productivity perspective we can establish using known practice, how long it should take to build a program or course based on what a quality program or course requires and the benchmarks for determining whether the work to do so was effective in practice. We know the process because we have a comparable and visible professionally controlled design and as a result can build proximal measures of productivity for learning and teaching at scale across a university. This has the potential to operationally define quality and link it directly to cost in ways that make it possible to bring greater precision to the development of courseware and differentiate the SOU university.

6. A Distinctive Student Growth Message: The SOU offers a university the opportunity to go beyond the rhetorical claims about learning and teaching success. The SOU can provide the distinctive differentiator that universities are seeking in the highly competitive contexts in which they operate. They can make an empirical case for their strengths and show those strengths as clear points of differentiation. Moreover a shared schema along with the transparency and understanding of learning and teaching derived from a SOU approach, can uniquely position students as strong advocates for their alma- maters. They will know what they have accomplished, how it happened, and how it is different.

References

ACT (2009). *ACT report: Voluntary system of accountability learning gains methodology.* Retrieved from https://cp-files.s3.amazonaws.com/22/ACTReport_LearningGainsMethodology.pdf

Bowker, G., & Star, S. (2000). *Classification and its consequences.* Cambridge, Massachusetts: The MIT Press.

Dvorak, N., & Busteed, B. (2015). It's hard to differentiate one higher-ed brand from another. *Gallup Business Journal.* Retrieved from http://www.gallup.com/businessjournal/184538/hard-differentiate-one-higher-brand.aspx

Harrison, L., & Mather, P. (2016). *Alternative solutions to higher education's challenges: An appreciative approach to reform.* New York: Routledge.

Hollands, F., & Tithali, D. (2014). MOOCS: Expectations and reality, center for benefit—cost studies of education, Teachers College Columbia University. Retrieved from http://www.academicpartnerships.com/sites/default/files/MOOCs_Expectations_and_Reality.pdf

Jankowski, N., Ikenberry, S., Kinzi, J., Kuh, G., Shenoy, G., & Baker, G. (2012). *Transparency & accountability: An evaluation of the VSA College portrait pilot.* Retrieved from http://www.learningoutcomeassessment.org/documents/VSA_001.pdf

Massey, W., Sullivan, T., & Mackie, C. (2012). Data needed for improving productivity measurement in higher education. *Research and Practice in Assessment, 7,* 5–15. Retrieved from http://www.rpajournal.com/dev/wp-content/uploads/2012/11/SF.pdf

Pascarelli, E., & Terenzini, P. (2005). *How college affects students (Volume 2): A third decade of research.* San Francisco: Jossey-Bass.

Springer, M. G., Ballou, D., Hamilton, L., Le, V., Lockwood, J. R., McCaffrey, D. et al. (2010). *Teacher pay for performance: Experimental evidence from the project on incentives in teaching.* Nashville, TN: National Center on Performance Incentives at Vanderbilt University. Retrieved from http://www.rand.org/content/dam/rand/pubs/reprints/2010/RAND_RP1416.pdf

Chapter 11
Engaging with Transformational Change

Abstract This chapter provides advice to leaders about engaging with transformational change including key guidance on assessing the organization's appetite for change, building capacity, managing the process, cultivating and sustaining support, and ensuring quality communication.

Keywords Transformational change · Regulatory framework · Human resource · Autogenetic · Project team · Middle managers · Executive leadership · Leadership structure · Divisions

The scope and depth of the SOU approach means that a university can get to the starting line with a design methodology that holds the requisite power and promise to create sustainable and scalable change in learning and teaching. Such an approach requires an appetite for change that is also commensurate with the scope and depth of the model.

There are a number of key factors a university needs to consider before embarking on transformational change in order to avoid inviting unnecessary stress and cost. For large complex organizations, the idea of transformational change is often more appealing than the heavy lifting required to make it happen. To engage with the SOU model first requires an understanding of what transformational change means. Is the organization prepared to look at its technological capability, its human resource model including role descriptions and promotion criteria, and its approach to leadership, its governance and regulatory framework? As noted earlier, there is an allure about a change initiative that does not create much change. Given the benchmarks for determining the success of educational initiatives are often ambiguous and almost always contested, it is altogether possible to talk up and market non-transformational change as if it was. Victory can be declared without disturbing the equilibrium of the organization. Under these circumstances it is not surprising that most of the significant change in the field of education has been done to it not by it (Tyack and Cuban 1995). This history of ambiguity, contestation and premature claims of "*victory*" makes educational organizations prime candidates for

© The Author(s) 2016
A. Bain and L. Zundans-Fraser, *Rising to the Challenge of Transforming Higher Education*, SpringerBriefs in Education,
DOI 10.1007/978-981-10-0261-8_11

claiming an appetite and desire for transformational change without actually fully appreciating what that means. An organization needs to be clear about the hard work of transforming itself.

A second consideration is to recognize the learning and teaching research base described here although fundamental to the field of education is not widely known or understood in higher education organizations. Individuals can and do function at all levels in universities leading learning and teaching without knowing much about it. This is not to indict or admonish, but to signal that a university that aspires to be a genuine learning organization will need as part of its plan to come to terms with this somewhat uncomfortable truth and reconcile role and capacity whereby leaders of learning and teaching will require leadership level knowledge and skill in learning and teaching. Making this alignment happen also needs to be part of the change appetite of the organization.

Related to two is identifying a project team with the capacity to help bridge the initial gap between the organization's current form and capacity and what it needs to become. Their role is not messianic although an initial injection of capacity is essential.

A fourth consideration pertains to the active and dynamic engagement with the change process by executive leadership and the peak governance body of the organization. Leadership needs to create the sense of urgency, build the working coalitions, and reconcile the change with the culture of the organization (Kotter 1998). While the outcome of the SOU is to create a bottom-up self-organizing system, change in human systems is rarely autogenetic. In the beginning, leadership must be catalysts for change leading from the front, mobilizing all levels of the organization, while knowing that they are leading a process to create a flatter, networked, less segmented and less hierarchical organization as a result of the SOU.

A fifth consideration involves a serious initial effort to cultivate the support of middle management who keep an organization functioning while change is happening. Individuals in these roles are critical to the ground level support for change, its communication and implementation. In change processes communication is frequently partitioned and segmented, assigned to the roles of offices or individuals. This is good for the production of newsletters, promotional materials and press releases. All important. However, the really critical communication occurs through the existing organizational structure and with those individuals who assume operational responsibility. Middle managers communicate with their constituencies because they are engaged—the change is always on their radar. Poor communication usually means it is not. The support of middle managers should not be assumed in the loosely coupled leadership structures of contemporary universities.

A sixth consideration is to reconcile scope with the capacity and appetite of the broader organization. A key theme of this brief has been to highlight that to change learning and teaching is to change the whole organization. This means that all of the ancillary divisions need to be part of the process. Divisions of information technology, human resources, learning and teaching, academic governance all need to be partnered with and valued in the process.

These six areas should not be confused with the full scope of considerations required for effective organizational change. They are considerations and assessments to be made prior to embarkation.

References

Kotter, J. (1998). Leading change: Why transformation efforts fail. In *Harvard business review on change* (pp. 1–20). Cambridge, MA: Harvard University Press.
Tyack, D., & Cuban, L. (1995). *Tinkering toward utopia: A century of public school reform.* Cambridge, MA: Harvard University Press.

Chapter 12
Conclusion

Abstract In this concluding chapter we describe four key considerations when using the self-organizing university approach and the importance of designing a process that is capable of delivering change at scale. The chapter includes a diagrammatic representation of the theory to practice trajectory of the model.

Keywords Theory · Transformative · Efficacy research · Organizational design · Policy · Governance · Roles · Emergent · Higher education

The body of work underpinning the SOU, conducted over two decades, has established a trajectory from the literature on theories of self-organization to the development of a set of design principles subsequently applied to organizational design in education. The design is also informed by over sixty years of efficacy research in learning and teaching that has been applied to those principles. Theory, learning and teaching research, efficacy research and higher education program design sets the foundation for the transformational redesign of a university. Figure 12.1 describes the research trajectory.

Four important considerations need to be emphasised in concluding this brief. First, the SOU is a transformative approach focused on change at a scale capable of addressing the challenges faced by universities and identified in the literature. Its focus is on the things that need to happen to design and deliver better learning and teaching at scale. In doing so, it makes extant connections among theory, research, and design. Second, the foundation of the model is expansive. The approach is underpinned by extensive, well-accepted longitudinal research about learning and teaching that sits at the heart of the design. The account of theory, learning, teaching and efficacy research provides a clear rationale for, and line of sight to, an organizational design that maps from theory to practice. This is especially the case when the challenges facing universities and the magnitude of the change and curricular problems we face are interrogated and understood. Third, understanding

© The Author(s) 2016   77
A. Bain and L. Zundans-Fraser, *Rising to the Challenge of Transforming
Higher Education*, SpringerBriefs in Education,
DOI 10.1007/978-981-10-0261-8_12

Fig. 12.1 Theory to research
and practice trajectory

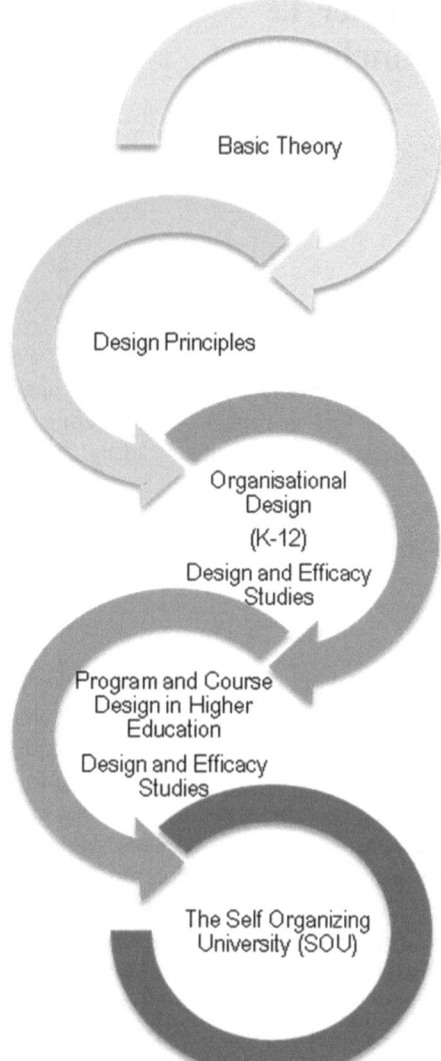

the SOU is about examining the connections in practice, looking at the detailed relationships among policies, tools, governance processes and roles. The design is an emergent expression of the theory and research described here. It can be "simplified up" (Drengenberg 2013), although most important is the way the contents of this brief are connected to the actual known practice of the SOU in higher education. Fourth, the SOU initiative creates the conditions to extend the research described here by investigating the effects of the SOU at scale. The continuation of a robust research agenda can evolve the model.

Reference

Drengenberg, N. (2013). *Joining the dots: Simplifying up*. Unpublished manuscript. Bathurst: Australia University.

Glossary

Constructive Alignment The components in the teaching system, especially the teaching methods used and the assessment tasks are aligned to the learning assumed in the intended outcomes.

Dispersed Control Empowering those with agency in the system, regardless of their status within an organization.

Edge Technology A model for developing technologies that improve the quality of learning and teaching.

Embedded Design A principle that involves thinking about and acting upon the ways the commitments can be enacted in an organization's design.

Emergent Feedback Feedback emerges all the time from normal work. Constant exchange among individuals and groups within an organization can create emergent change.

Feedback Program team members use the feedback capabilities of TAES to self-evaluate, as well as seek the perspectives of peers, external advisors, and university leadership about key facets of the program design process. A set of research-informed feedback questions are connected to each module of TAES.

Learning Analytics Tools that enable the analysis of a learning experience from conceptualization to student performance throughout a program.

Learning Experience A learning experience is the activity that enacts the learning outcomes of a course and makes completion of an assessment task possible.

Module Module refers to a part of the program review, design and accreditation process in TAES.

Program A program is a coherent collection of courses that make up a recognised final award or degree for example, Bachelor Degree, Graduate Diploma.

Program Space Is a design space created in TAES for a given program that includes program conceptualization, standards mapping, program outcomes, assessment tasks, course and unit design.

© The Author(s) 2016
A. Bain and L. Zundans-Fraser, *Rising to the Challenge of Transforming Higher Education*, SpringerBriefs in Education,
DOI 10.1007/978-981-10-0261-8

Program Team All members of the faculty involved in designing the program and course material in TAES.

Schema Provides the form required for shared understandings, the creation of a genuine community of practice, and the flexibility to be dynamic and subject to change based upon emergent feedback.

Self-Organization An explanatory framework for systems in nature and those involving human action and intervention. The role and contribution of self-organization extends beyond the metaphorical through the application of six design principles—simple rules, embedded design, similarity at scale, emergent feedback, dispersed control and schema.

Self-Organizing University (SOU) A transformational change model for learning and teaching in higher education.

Similarity at Scale In higher education, similarity at scale is achieved by ensuring that the tools, systems and methods that operationalize the theory are similar at all levels of the organization.

Simple Rules or Commitments This principle determine the values, beliefs and dispositions that a program team believe are a necessary pre-condition for action.

Tools for Adaptive Education Systems (TAES) A suite of technology tools for program design, delivery and analytics in higher education.